ACID PROTEASES
Structure, Function, and Biology

ADVANCES IN EXPERIMENTAL MEDICINE AND BIOLOGY

Recent Volumes in this Series

ACID PROTEASES
Structure, Function, and Biology

Edited by

Jordan Tang
Oklahoma Medical Research Foundation

SPRINGER SCIENCE+BUSINESS MEDIA, LLC

Library of Congress Cataloging in Publication Data

Main entry under title:

Acid proteases.

(Advances in experimental medicine and biology, v. 95)
"Proceedings of a conference on acid proteases: structure, function, and biology, held at the University of Oklahoma, Norman, Oklahoma, November 21-24, 1976."
Includes index.
1. Protease—Congresses. I. Tang, Jordan.
QP609.P7A24   574.1'9256   77-13032
ISBN 978-1-4757-0721-2   ISBN 978-1-4757-0719-9 (eBook)
DOI 10.1007/978-1-4757-0719-9

Proceedings of a Conference on Acid Proteases: Structurt,
Function, and Biology held at the University of Oklahoma,
Norman, Oklahoma, November 21—24, 1976

©1977 Springer Science+Business Media New York
Originally published by Plenum Press, New York in 1977

PREFACE

In the past ten years, a number of proceedings of symposia on the structure and function of proteolytic enzymes have been published. Their coverage of acid proteases has been limited, mainly due to the lack of significant new information on the structure of these enzymes. In the last four years, however, the primary and tertiary structures of a number of acid proteases have been determined, prompting the need to discuss the meanings of the old data and the possibilities for new experimentations. It was for this purpose that the "Conference on Acid Proteases: Structure, Function, and Biology" was organized. It took place at the University of Oklahoma on November 21-24, 1976. This book is a collection of the main lectures delivered at the Conference.

Acid Proteases, by definition refers to a group of proteases having an optimal pH in acidic solutions. The classic examples are pepsin and chymosin. Some catalytic features are obviously shared by these proteases, most notably, their inhibition by pepstatin. The use of active center-directed inactivators such as diazoacetyl-norleucine methyl ester and 1,2-epoxy-3-(p-nitrophenoxy)propane has shown that two catalytic aspartyl residues are present in most of these enzymes. These apparent common features have prompted the suggestion by several investigators to name this group of enzymes "aspartyl proteases" or "carboxyl proteases". Such proposals are particularly valid if one considers that the optimal pH of renin is about 6, but its catalytic residues and mechanism obviously belong to that of the acid proteases. Regardless of the name eventually adopted, there is little question that this is a group of proteases with a structure-function relationship different from other groups of proteases. They appear to have some important functions in various biological systems.

It is my hope that the information collected in this volume will stimulate a broader interest in future investigations of acid proteases.

For the Conference, I wish to acknowledge the sponsorship of the Oklahoma Medical Research Foundation, and financial support through grants from the National Institutes of Health (GM-23661) and from the National Science Foundation (PCM76-17344).

Special thanks should go to my colleague Dr. Jean A. Hartsuck who has contributed thoughts and work, both for the Conference and for the editing of the Proceedings, to Dr. David Davies of NIH and Dr. Tadashi Inagami, Vanderbilt University, for help in designing the scientific program, to Mrs. Ester Pahlka, Conference Secretary, for able organizational planning and execution, to Mr. and Mrs. Milton Smith, Mr. David Ponder, Ms. Jan Rogers, Ms. Mary Simpson, Ms. Dana Pitts, Ms. Vicki Cassady, Ms. Beverly Yurtis, and other volunteers from the Oklahoma Medical Research Foundation, for working at the Conference with great enthusiasm, and to Ms. Lois Fagin, for help in editing these Proceedings.

                                                        Jordan Tang

CONTENTS

PRIMARY AND THREE-DIMENSIONAL STRUCTURE

MECHANISM OF PEPSINOGEN ACTIVATION

CATALYTIC MECHANISM OF PEPSIN

ACID PROTEASES IN VARIOUS BIOLOGICAL SYSTEMS

PRIMARY AND THREE-DIMENSIONAL STRUCTURES

COMPARISON OF THE PRIMARY STRUCTURES OF ACIDIC PROTEINASES AND OF THEIR ZYMOGENS

Bent Foltmann and Vibeke Barkholt Pedersen

Institute of Biochemical Genetics, University of Copenhagen
Ø.Farimagsgade 2A, DK-1353 Copenhagen K, Denmark

Proteinases with a pH-optimum of general proteolytic activity at pH below 6 are called acidic proteinases. Such enzymes were at first recognized as the proteinases of the vertebrate gastric juice, and for many years porcine pepsin and calf chymosin have been the most well-known members of this group.

Pepsin, one of the first enzymes to be discovered (1), was the second to be crystallized (2), and has been the subject of a large number of investigations (3,4).

The use of chymosin in cheesemaking has been known long before the terms "ferment" or "enzyme" were conceived. It was also one of the first enzymes to be named (5). Extensive literature exists, especially about the milk clotting process; the biochemistry of chymosin has been reviewed by one of the authors (6).

Despite much investigation, little was known until recently about the relationship between structure and function of pepsin and chymosin; only about ten years ago (7) did we begin to understand that they are representatives of a large group of enzymes.

Other chapters of this book offer a detailed account of the catalytic mechanism of the acidic proteinases; here it will suffice to briefly mention that most of the acidic proteinases have two especially reactive aspartate groups (8). One can be labelled with active site-directed diazo compounds (9,10), and another can be labelled with active site-directed epoxy compounds (11). Both types of labelling will inactivate the enzymes. All such enzymes are also inhibited by pepstatin, a peptide produced by strains of Streptomyces (12).

3

Today it is generally assumed that all enzymes inactivated by these inhibitors share a common catalytic mechanism. The terms "aspartate proteinases" or "carboxyl proteinases" have been suggested for these enzymes. Not all proteinases with low pH-optimum belong to this group. For instance, proteinases from Aspergillus niger (12) and Scytalidium lignicolum (13) are insensitive to these inhibitors. We shall limit this discussion to acidic proteinases with aspartic acid residues in the active site. The term "aspartate proteinases" (14) in the same manner as serine proteinases is used.

VERTEBRATE GASTRIC PROTEINASES

The gastric proteinases form a group with many members; before discussing their structure it is appropriate to briefly discuss the classification of and terminology for these enzymes. Unfortunately, no common system of classification has been used, and chemically, the individual components have in most cases been poorly characterized. Thus, it is difficult to evaluate the relation between components of gastric proteinases described by different authors. Additionally, when comparing the properties one must remember that most investigations on the gastric proteinases have been carried out with the commercially available porcine pepsin, and in many textbooks, results of these investigations are sited as characteristic for all pepsins disregarding the possible differences between the properties of pepsins from different species.

From porcine gastric mucosa and crude preparations of porcine pepsin, Ryle and collaborators prepared a main component, pepsin A, and three minor components called pepsin B, C, and D (15). Human pepsinogens and pepsins have been fractionated by chromatography on DEAE-cellulose (16,17,18). In these investigations the individual components are designated with Roman numerals according to the order of elution. A similar designation has been used by Antonini and Ribadeau Dumas (19) in their study on bovine pepsinogens. Tang and co-workers isolated pepsin and a second component called gastricsin from human gastric juice, by chromatography on Amberlite IRC-50 (20), and from crude commercial porcine pepsin.

Other groups of biochemists have numbered the individual components of human pepsinogen according to their decreasing electrophoretic mobilities at pH 8.5 (21,22). The pepsins have likewise been numbered by their decreasing electrophoretic mobilities at pH 5.0. Etherington and Taylor (23) and Mangla et al. (24) showed that the relative electrophoretic mobilities of pepsinogens were retained in the electrophoretic mobilities of the resulting pepsins. Shaw and Wright (25) obtained similar results by analysis of pepsinogens from cat gastric mucosa and the pepsins derived from them.

If the human gastric proteinases from adult subjects are analyzed by immunochemical methods, they may be classified in two main
groups (26,27,28), one of which is predominant and secreted into the
fundus, while the minor components are also formed in the antropyloric part of the stomach. In addition to these components, Hirsh-
Marie et al. (29) have also demonstrated the presence of a human
fetal pepsin. The observations on the immunochemical characteristics
of human gastric proteinases resemble results in studies of bovine
proteinases where two pepsins and chymosin were all found to be
immunochemically independent (30).

It may be premature to make generalizations based on fragmentary
information from different sources, but despite this we believe we
discern a general pattern, and propose the following classification.

In mammalia, the zymogens of the gastric proteinases and the
resulting active enzymes may be divided into three main classes:

Class 1: The main component of the pepsin group, pepsin A (EC 3.4.
23.1).

The enzymes are predominantly secreted in the fundus. The
general proteolytic activity shows an optimum of about pH 1.8 to 2.2.
Low activity towards hydrolysis of carbobenzoxy-tyrosyl-alanine and
high activity towards acetyl-phenylalanyl-diiodotyrosine has been
reported as characteristic for pepsin A (20). By electrophoresis at
pH 8.5 for pepsinogen, or pH 5.0 to 5.4 for pepsin, this group of
enzymes has the greatest mobility toward the anode (21-25).

The complete amino acid sequence of porcine pepsinogen A has
been determined; partial structures of bovine pepsinogen A and of
human pepsin are known.

Class 2: The minor components of the pepsin group.

Some of these enzymes are mainly secreted in the antropyloric
part of the stomach. Names like pepsin B, C, gastricsin, or pyloric
pepsin have been used for members of this group. Interspecies
relationship between these enzymes is not yet clear. The general
proteolytic activity shows a maximum of about pH 3.0 (17,19). Low
activity towards hydrolysis of acetyl-phenylalanyl-diiodotyrosine
and high activity towards carbobenzoxy-tyrosyl-alanine have been
reported as characteristic for both pepsin C and gastricsin (20);
the two designations most likely describe the same enzyme. A
similar relation between the activities toward the two substrates
has been found for bovine pepsin B (M. K. Harboe, personal communication).

Partial structures of human gastricsin and bovine pepsinogen B
have been determined.

Class 3: Fetal or neonatal pepsins.

Zymogens or enzymes belonging to this class have been found in humans (29), rats (31), rabbits (32), lambs (33), and calves. For historic and technologic reasons the names chymosin and rennin have been used for the enzyme from calves. The name chymosin will be used in this paper. By electrophoresis these enzymes belong to the slowly moving components.

The complete primary structure of bovine prochymosin B has been determined.

Within each of the main three classes, several factors may contribute to further heterogeneity:

(a) Multiple gene products.

Table I shows examples of single or few amino acid substitutions, e.g. chymosin A and B.

(b) Different degrees of phosphorylation.

Porcine pepsin D is unphosphorylated pepsin A (34). In bovine pepsin A, four components with different degrees of phosphorylation have been observed (35).

(c) Autolysis.

Different NH_2-terminals may be produced during activation (36); chymosin C is a partial degradation product (37).

(d) Carbohydrates.

A content of carbohydrates is a theoretical possibility for heterogeneity in the gastric proteinases. Kageyama and Takahashi (37a) have shown that the carbohydrates are linked to an asparagine residue in the pepsin part of the monkey pepsinogen. At least two carbohydrate containing residues have been observed in the sequence of human pepsin (Sepulveda and Tang, personal communication). On the other hand, Stepanov et al. (38) have reported that the content of carbohydrate in chromatographically purified preparations of porcine pepsinogen is negligible, and we have not been able to demonstrate the presence of carbohydrate in any of the bovine gastric zymogens (Foltmann et al. unpublished observations). The extent of carbohydrate contribution to heterogeneity of these proteases is still uncertain.

Several pepsinogens and pepsins have been reported in dogfish and chickens, but we do not find the information sufficient to classify these enzymes in the general scheme suggested above.

OTHER ACIDIC PROTEINASES FROM VERTEBRATES

In addition to the gastric proteinases, several other vertebrate proteinases are inhibited by the diazo compounds, by pepstatin, or by the epoxy compounds. Among these are renin (39,40), cathepsin D (41,42), thyroid proteinase (43) and an acidic seminal proteinase (44). Even though renin does not have a low pH-optimum, its inhibition pattern indicates that it along with the above mentioned enzymes, belongs to the aspartate proteinases.

No sequence studies on these enzymes have been published.

MICROBIAL PROTEINASES

A series of microbial proteases useful in food production, especially as rennet substitutes, has been isolated (45). Many of these have an optimum pH below 6 and are inhibited by specific pepsin inhibitors described above.

The sequence of penicillopepsin from Penicillium jantinellum is almost completed, while partial sequences are known for Rhizopus-pepsin (Rhizopus chinensis) and Mucor miehei proteinase. A proteinase from Endothia parasitica and the enzymes from P. janti-nellum and R. chinensis have been solved for x-ray crystal structure (see Chapter 3, 4, and 5).

PRIMARY STRUCTURES

When comparing the primary structures of a series of related peptide chains, it is common to align the individual chains with gaps to maximize homology. With the fragmentary information available about the amino acid sequences of many aspartate proteinases, it is difficult to make a final recommendation about the numbering of amino acid residues, but the following facts must be considered:

The aspartate proteinases of the gastric juice are all secreted as inactive precursors. These precursors are irreversibly converted into active enzymes during which process NH_2-terminal segments (activation segments or proparts) of the peptide chains are released. Renin and the seminal aspartate proteinase are also reported to be secreted as precursors (44,46, also see Chapters 14, and 20 in this volume), while no zymogens for microbial aspartate proteinases have been found.

Table I

Alignment of amino acid sequences of acidic proteinases*

The sequences are expressed in the single letter code according to the recommendation of IUPAC/IUB.

A: Ala, B: Asx, C: Cys, D: Asp, E: Glu, F: Phe, G: Gly, H: His, I: Ile, K: Lys, L: Leu, M: Met, N: Asn, P: Pro, Q: Gln, R: Arg, S: Ser, T: Thr, V: Val, W: Trp, X: unknown, Y: Tyr, Z: Glx. -: gap, /: end of known sequence or end of peptide. Period shows that overlap is inferred from homology. Commas indicate residues present from amino acid composition.

S-S bridges: C(91) to C(96), C(252) to C(256), and C(296) to C(329).

Where zymogens have been sequenced, the abbreviations of the zymogens are preceding lines containing proparts of the structure. After residue 77, all structures are named by abbreviated form of the active enzyme. Note that the name of a given enzyme is only shown in the sections of the table where sequences are known. The two lower lines illustrate residues common in gastric enzymes and common in all sequenced structures.

The body of Table I is shown on the next 5 pages

```
                              10              20              30           40
Pig p.gen A     L V K V P L V R K K S L R Q N L I K D G K L K D F L K T H K H N P A S K Y F
Ox p.gen A      S V V K I P L V K K K S L R Q N L I E N G K L K E F M R T H K H N P A E K Y I
Ox p.gen B      L V K I P L K K F K S I R E I M K E K G L L Z B F L R T Y K H N P A E K Y -
Calf prochym.   A E I T R I P L Y K G K S L R K A L K E H G L L E D F L Q K Q Q Y G I S S K Y S
Seal p.gen      A I I R V P L V K K K A L R K B L R Z H G L L Z B F/
Chicken p.gen   S I H R V P L L K G K.S.L R K Q L.K D H G L L E D F/
Dogfish p.gen   L I R V P L I K/

Common              P L     K       R       G   L       F             K Y


                              50              60              70
Pig p.gen A     P E A A A L I G D E P L E N Y L - D T E Y F - - G T I G T P A Q D F T V I F
Hum. pep. A       V D E Q P L E N Y L - D M E Y F - - G T I G I G T P A Q D F T V V F
Hum. pep. C       S V T Y E P M A - Y M - D A A Y F - - G E I S I G T P A Q N F X V/
                  S V
Ox p.gen A      R E A A T L V S E Q P L Q N Y L - D T E Y F - - G T I G I G T P A Q D F T V I F
Ox p.gen B      R F G D F I V A T E P M - D Y/   /F - - G Z I S I G T P P Z B F/
Calf prochym.   - - G F G E V A S V P L T N Y L - D S Q Y F - - G K I Y L G T P P Q E F T V L F
Penic. pep.     A A S G V A T N T P T A N - - D E E Y I T P V T I G - G T - I - L N L N F
Rhiz. pep.        V G T V P M T D Y G N D I E Y Y - - G Q V T I G T P G/
M. miehei p.    A A A D G S V D T P - - G Y Y - D F D L/

Common gast.            P     Y     D     Y F     G     I   G T P   Q   F T V   F
Common all                    D                             G T                   F
```

Table 1 (continued)

```
              80              90             100            110
Pig pep. A    D T G S S N L W V P S V Y C S - S L A C S D H N Q F N P D D S S T F E A T S Q E
Hum. pep. A   D T G /
Ox pep. A     D T G S S N L W V P S I Y C S - S E A C T N H N R F N P Q D S S T Y E /
Calf chym.    D T G S S D F W V P S I Y C K - S N A C K N H Q R F D P R K S S T F Q N L G K P
Penic. pep.   D T G S A D L W V F S T E L P A S - Q Q S G H S V Y N P S A T G K - E A S G Y T
M. miehei p.                        / K G C T K S E G C V G S R F F D P S T S /

Common gast.  D T G S S       W V P S   Y C   S       A C     H         F     P   S S T
Common all    D T G S         W V       S             S                       P


              120            130            140            150
Pig pep. A    L S I T Y G T G S M - T G I L G Y D T V Q V G G I S D T N Q I F G L S E T E P G
Ox pep. A              / M - T G /                                      / L S E T E P G
Calf chym.    L S I H Y G T G S M - Q G I L G Y D T V T V S N I V D I Q Q T V G L S T Q E P G
Penic. pep.   W S I S Y G D G S S A S G N V F T D S V T V G G V T A H G Q A V E A A Q Q I S A

Common gast.  L S I   Y G T G S M       G I L G Y D T V     V       I D       Q   G L S   E P G
Common all    S I     Y G   G S         G         D   V     V                 Q             Q
```

```
                    160         170          180          190
Pig pep. A    S F L Y Y A P F D G I L G L A Y P S I S A S G A T P V F D N L W D Q G L V S Q D
Hum. pep. C              / G L A Y P A L S V D E A T T A M /   / I W D Q G.L V S Z B
Ox pep. A     S F/
Ox pep. B                / A Y P S L S V B G A T T V L Z Z M /
Calf chym.    D V F T Y A E F D G I L G M A Y P S L A S E Y S I P V F D N M M N R H L V A Q D
Penic. pep.   Q F Q Q D T N N D G L L G L A F S S I N - T V Q P S Q T T F F D T T V K S S L A
M. miehei p.                                      / N T N S G T G E V F G G V N N T L L S G D I

Common gast.              Y A   F D G I L G   A Y P
Common all                D G   L G   A                                                  L V
```

```
                    200         210          220          230
Pig pep. A    L F S V Y L S S N D D S G S V V L L G G I D S S Y Y T G S L N W V P V - S V E G
Ox pep. A     L/        / L S S N E E S G/                                     / L N W V P/
Calf chym.    L F S V Y M D R D G Q E - S M L T L G A I D P S Y Y T G S L H W V P V - T V Q Q
Penic. pep.   Q P L F A V A L K H Q Q P G V Y D F G F I D S S K Y T G S L T Y T G V D N S Q G
M. miehei p.  A Y T D V M/

Common gast.  L F S V Y                           L G   I D   S Y Y T G S L   W V P V       V
Common all    L F S V                             G     I D   S   Y T G S L                 V
```

Table 1 (continued)

```
                     240         250         260         270
Pig pep. A     Y W Q I T L D S I T M D G E T I A C S G G C Q A I V D T G T S L L T G P T S A I
Hum. pep. A                      /D G E A I A X S E G X E A X V D T/      /L T G P T S P I
Ox pep. A      /Y W Q/   /I T M D G E S I . A C . S . N G C E A . I V D T G T S L/
Calf chym.     Y W Q F T V D S V T I S G V V V A C E G G C Q A I L D T G T S K L V G P S S D I
Penic. pep.    F W S F N V D S Y T A G S Q S G L G F - - S G I A D T G T T L L L B D S V V
M. miehei P.                                                                /P S S A A

Common gast.   Y W Q         T D S         T     G     A C     G C A I   D T G T S   L   G P   S I
Common all         W         D S         T                       I       D T G T     L           S
```

```
                     280         290         300         310
Pig  pep. A    A I N I Q S D I G A - S E N S D G E M V I S C S S I D S L P D I V F T I N G V Q
Hum. pep. A    A - N I Q S D I G A - S E N S D G E M V V S A I X S L P D I V F T I X G V E
Ox pep. A                                                        /I N S L P D/
Calf chym.     L - N I Q Q A I G A - T Q N Q Y D E F D I D C D N L S Y M P T V V F E I N G K M
                                           G
Penic. pep.    S Q Y Y S Q V S G A Q Q D S N A G G Y V F X C S/                        /T A T
M. miehei P.   S K I V K A S L P B - A T Z T Z Z G,W,V V P C A S Y Q N S K S T I S I V M/

Common gast.       N I Q       I G A     N                   C         P     V F I       G
Common all         N I Q       I G A                         C
```

```
                      320                       330                 340               350
Pig pep. A     Y P L S P S A Y I L Q D D D S - C T S G F E G M D V P T S S G E L W I L G D V F
Hum. pep. A    Y P/                                    /E G M N L P T E X G E L/ / I L G D V F
Ox pep. A                                                             /L W . I L G D V F
Calf chym.     Y P L T P S A Y T S Q D Q G F - C T S G F Q S E N - - - H S - Q K W I L G D V F
Penic. pep.    V P G S L I N Y G P S G N G S T C L G G I Q S N/              /G D I F
M. miehei p.                      /L L P V D Q S N E - T C M F/

Common gast.   Y P L       P S A Y       Q D   C T S G F                W I L G D V F
Common all     P                                   C   G                    G D   F


                      360                       370
Pig pep. A     I R Q Y Y T V F D R A N N K V G L A P V A
Hum. pep. A    I R Q Y F T V F D R A N N Q V G L A P V A
Hum. pep. C     /Q F Y T V F D R A N N K E G L A P V A
Ox pep. A      I R Q Y F T V F D R G N N Q I G L A P V A
Calf chym.     I R E Y Y S V F D R A N N L V G L A K A I
Penic. pep.    L K S Q Y V V F D S D G P Q L G F A P Q A

Common gast.   I R     V F D R   N N   G L A
Common all             V F D         G   A
```

*Footnote for Table I, Source of Information:

Porcine pepsinogen and porcine pepsin A (porc.p.gen A, porc. pep. A). Residues 3-41 (47), residues 42-46 (48), residues 47-373 (49,50). The results are confirmed by the partial structures of Stepanov et al. (51,52).

The main component of active pepsin has Ile-47 as NH_2-terminus; minor amounts occur with Ala-45 as NH_2-terminus. Ser-114 is phosphorylated; Ile-276 was not present in the pepsin analyzed by Morávek and Kostka (50); the residue was partly present in preparations analyzed by Sepulveda et al. (49). At position 301, glutamine was reported in minor amounts by Sepulveda et al. (49). Residue 310 was determined by Morávek and Kostka as asparagine and as aspartic acid by Sepulveda et al.

Human pepsin A (Hum. pep. A)
Residues 47-80 (53), residues 347-373 (54). Other sequences: Personal communication, P. Sepulveda and J. Tang (Oklahoma Med. Res. Found.).

Human pepsin C = Human gastricsin (Hum. pep. C)
Residues 46-70 (53), residues 355-373 (54). Other sequences: Personal communications, P. Sepulveda and J. Tang (Oklahoma Med. Res. Found.).

Bovine pepsinogen A, bovine pepsin A (bov. p.gen A, bov. pep. A)
Residues 2-46 (55), residues 47-111 (56), residues 259-265 (57), residues 355-373 (58). Other sequences: Personal communication, M. K. Harboe (Inst. Biochem. Gen., Univ. Copenhagen). NH_2-terminus of the active enzyme is Val-47.

Bovine pepsinogen B, bovine pepsin B (bov. p.gen B, bov. pep. B)
Residues 2-27 (59). Other sequences: Personal communication, M. K. Harboe (Inst. Biochem. Gen., Univ. Copenhagen). NH_2-terminus of the active enzyme is Ile-46.

Calf prochymosin, calf chymosin (calf prochym., calf chym.).
NH_2-terminus of the active enzyme is Gly-45. Residues 1-105 (60,61). S-S bridges and residues 357-373 (7). The rest of the sequence from Foltmann and coworkers unpublished results. Residue 290 is aspartic acid in chymosin A and glycine in chymosin B.

Seal pepsinogen (Seal p.gen)
Residues 2-27 (59).

Chicken pepsinogen (chicken p.gen).
Residues 2-27, personal communication V. Kostka and H. Keilova (Inst. Org. Chem. Biochem. Czech. Acad. Science, Prague).

Dogfish pepsinogen (Dogfish p.gen).
Residues 2-10 (59).

Penicillopepsin (penic. pep.)
Residues 43-243, and 349-373 (62). Other sequences: Personal
communication, T. Hofmann (Dept. Biochem., Univ. Toronto).

Rhizopus-pepsin (Rhiz. pep.)
Residues 47-70 (53). Other sequences: Personal communications,
P. Sepulveda and J. Tang (Oklahoma Med. Res. Found.).

Mucor miehei proteinase (M.miehei p.).
All sequences: Personal communication, A. M. Faerce (Inst. Biochem.
Gen., Univ. Copenhagen). Identities with the other structures are
few and alignments must be considered with reservation. Carbohydrate
is attached to Asn-187.

(End of footnote for Table I)

The question is further complicated, as neither the zymogens nor the active enzymes shared a common position of the NH_2-terminus to serve as a starting point for the numbering.

Having considered these difficulties, we have concluded that for the present purpose it is most rational to number the residues from the NH_2-terminus of the longest known polypeptide chain of the gastric zymogens (prochymosin) and then continue by counting the longest completely sequenced chain of the active enzymes (porcine pepsin).

With these considerations in mind, we have aligned all available sequences in Table I.

DISCUSSION

The alignment in Table I includes sequences from 12 different zymogens and enzymes, only two of which are known completely and a third is almost completed. Despite the fragmentary data, we are able to deduct some general structure-function relationships of the aspartate proteinases and of the gastric zymogens.

Throughout the 373 positions of amino acid residues in the gastric zymogens, 175 are found with identical residues in the presently known structures. This number will probably decrease somewhat when more structures are determined. If we compare the gastric enzymes with the microbial enzymes represented by penicillo-pepsin, and Mucor miehei proteinase, 63 positions are identical.

These identities are not randomly distributed, but are clustered in groups, probably reflecting special functions of these sections of the peptide chains.

The first section with a highly homologous structure is in the zymogens between residues 5 and 15. Only four residues, 7, 8, 12 and 15, are identical in all seven zymogens, but a closer inspection shows that all the apolar and basic amino acid residues are highly conservative.

In 1966 it was suggested (6) that prochymosin and pepsinogen at neutral pH stabilizes in inactive conformation through electro-static interactions between positive charges of the active segment and negative charges in the enzyme part of the molecule. The sequence results are consistent with this hypothesis. Through the evolution from cartilaginous fishes, to birds and to higher vertebrates, the distribution of basic amino acids in the NH_2-termi-nal part of the zymogens has been conserved, and probably reflecting the functional significance of these groups.

The NH$_2$-termini of the active enzymes vary from position 43 (for Mucor miehei proteinase) to position 47 (for most of the pepsins). The peptide chains adjacent to the NH$_2$-termini show little homology, and the next cluster of highly homologous residues are found from Phe-77 to Ser-88. Here 8 residues of 12 are identical, including Asp-78 (residue 32 in the pepsin numbering of the active enzymes), the residue that has a specific reactivity towards epoxide inactivators.

From Ser-118 to Ser-125, from Asp-164 to Ala-170, and from Gly-214 to Leu-225 we observe 6, 5 and 9 identities out of 8, 7 and 12 positions respectively. The possible significance of these structural features is still unknown.

The aspartic acid residue 261 (215 in the pepsin numbering) is the site of esterification by diazoacetyl-norleucinemethyl ester or similar compounds. As might be expected for an active center residue, it is embedded in conservative segments from Ile-259 to Leu-267 where 6 identities are observed in 9 positions. Finally, in the COOH-terminal sequence from Gly-349 to Ala-370 eight residues are in identical positions. In this section Arg-354 and Arg-362 are common to all the gastric enzymes, while penicillopepsin has a lysine residue at position 354, and a serine residue at position 362.

Overall, about two-thirds of the identities are found in sections covering about 20% of the peptide chain. Two of these sections include the aspartate residues that have been proposed to participate in the catalytic mechanism. A possible functional importance of the other sections is not yet known, but model building based on the x-ray crystallographic data may reveal the contribution of some of these sections to the stability of the tertiary structures of these enzymes.

The results shown in Table I and this discussion indicate that the gastric proteases form a family of enzymes, closely related structurally. It also appears that pepsin A from different species (pig, ox and humans) are more closely related to each other than pepsin A is to other proteinases of the same species (e.g. bovine pepsin A, B, and chymosin); but with our present knowledge it is difficult to determine the interspecies relationship between bovine pepsin B and human pepsin C.

The penicillopepsin and Rhizopus-pepsin belong to the same family of enzymes, but as one could expect from the phylogenetic distance between vertebrates and fungi, the microbial enzymes show only a limited similarity to the gastric enzymes. From Mucor miehei proteinase no fragments corresponding to the highly homologous sections have as yet been sequenced. With the alignment outlined in

Table I, 20% of the residues in Mucor miehei proteinase are
identical to those of at least one other sequence of acidic proteinases. Thus as a working hypothesis we suggest that Mucor miehei
proteinase belongs to the same family.

If one compares the highly homologous amino acid sequences
around Asp-78 and Asp-261 a striking similarity is observed. As
pointed out by Sepulveda et al. (49), this may reflect an internal
gene duplication. However, the sequence Trp-Val-Pro also occurs
twice in the peptide chain of the gastric proteases, and if we
extend the comparison of internally repeated sequences to include
the residues from 221 to 229 we arrive at the pattern shown in
Table II. This alignment may reflect a gene multiplication in the
evolution of an ancestral gene for the aspartate proteinases and
this hypothesis simply accounts for the origin of all repeated
sequences in the peptide chain.

With our present knowledge it is very difficult to evaluate
the probability of the internal homologies. We are fully aware of
the risk in inventing homologies by suggesting a sufficient number
of gaps, but despite this we think that the internal homology suggested in Table II is an interesting possibility.

The tryptophan and the proline residues are not present in two
of the penicillopepsin sequences, and one may ask why traces of an
original structure apparently are retained more conservatively in
the gastric enzymes of the vertebrates than in penicillopepsin. The
answer could be that the advent of the zymogens has imposed a restriction on the evolution of the active enzymes which now, together
with the activation segments, have to fit into an inactive conformation. Thus it is feasible that some parts of the gastric proteases
are more conservative than corresponding parts of the fungal proteases.

Whether these speculations are true or not, we may conclude by
saying that the story of the acidic proteinases told at the molecular
level has just begun. We are convinced that in the near future many
interesting perspectives will be elucidated, allowing us a better
understanding of the general properties of this family of enzymes.

ACKNOWLEDGMENT

V. Kostka, H. Keilova, P. Sepulveda, J. Tang, T. Hofmann,
M. K. Harboe, and A.-M. Faerch have kindly made their unpublished
results available to us. We are greatly indebted to all.

Our own research has been supported by the Danish Natural
Science Research Council and by the Carlsberg Foundation.

Table II

Repeated sequence of acid proteinases
arranged in a possible internal homology

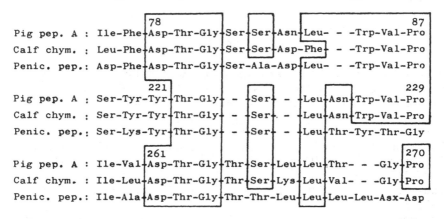

```
                      78                                        87
Pig pep. A :  Ile-Phe Asp-Thr-Gly Ser Ser Asn Leu-  - -Trp-Val-Pro
Calf chym. :  Leu-Phe Asp-Thr-Gly Ser Ser Asp-Phe   - -Trp-Val-Pro
Penic. pep.:  Asp-Phe Asp-Thr-Gly Ser-Ala-Asp Leu-  - -Trp-Val-Pro

                      221                                      229
Pig pep. A :  Ser-Tyr-Tyr Thr-Gly  -  Ser  -  Leu Asn Trp-Val-Pro
Calf chym. :  Ser-Tyr-Tyr Thr-Gly  -  Ser  -  Leu Asn Trp-Val-Pro
Penic. pep.:  Ser-Lys-Tyr Thr-Gly  -  Ser  -  Leu Thr-Tyr-Thr-Gly

                      261                                      270
Pig pep. A :  Ile-Val Asp-Thr-Gly Thr Ser Leu Leu Thr-  - -Gly Pro
Calf chym. :  Ile-Leu Asp-Thr-Gly Thr Ser Lys Leu Val-  - -Gly Pro
Penic. pep.:  Ile-Ala Asp-Thr-Gly Thr-Thr-Leu Leu Leu-Leu-Asx-Asp
```

SUMMARY

From the scattered information about primary structures of aspartate proteinases the following general features appear:

1) Sequence determinations show that two catalytically active aspartic acid residues are located in highly conservative surroundings.

2) Zymogens have so far only been found for extracellular aspartate proteinases of the vertebrates. The zymogens from the gastric mucosa are converted into active enzymes by a limited proteolysis releasing 42 to 44 residues from the NH_2-terminus. A common pattern of basic and apolar residues is observed in this NH_2-terminal segment.

3) In the presently known structures gastric proteinases and their zymogens have about 40% of all residues in common. The sequence of penicillopepsin shows 18% of identity with the gastric proteinases.

REFERENCES

1. Schwann, T. (1836) Mullers Arch. f. Anat. Physiol. 90-138
2. Northrop, J. H. (1930) J. Gen. Physiol. 13, 767-780
3. Fruton, J. S. (1970) Adv. Enzymol. 33, 401-444
4. Fruton, J. S. (1976) Adv. Enzymol. 44, 1-44
5. Deschamps (1840) J. de Pharm. Sci. Acces. 26, 412-420
6. Foltmann, B. (1966) C. R. Trav. Lab. Carlsberg 35, 143-231
7. Foltmann, B., and Hartley, B. S. (1967) Biochem. J. 104, 1064-1074
8. Knowles, J. R. (1970) Phil. Trans. R. Soc. London, Ser. B 257, 135-146
9. Rajagopalan, T. G., Stein, W. H., and Moore, S. (1966) J. Biol. Chem. 241, 4295-4297
10. Knowles, J. R., and Wybrandt, G. B. (1966) FEBS Lett. 1, 211-212
11. Chen, K. C. S., and Tang, J. (1972) J. Biol. Chem. 247, 2566-2574
12. Aoyagi, T., Kunimoto, S., Morishima, H., Takeuchi, T., and Umezawa, H. (1971) J. Antibiot. 24, 687-694
13. Murao, S., Oda, K., and Matsushita, Y. (1973) Agr. Biol. Chem. 37, 1417-1421
14. Kovaleva, G. G., Shimanskaya, M. P., and Stepanov, V. M. (1972) Biochem. Biophys. Res. Commun. 49, 1075-1081
15. Ryle, A. P. (1970) in Methods of Enzymology (Colowick, S. P., and Kaplan, N. O., eds.) 19, 319-336
16. Seijffers, M. J., Segal, H. L., and Miller, L. L. (1963) Amer. J. Physiol. 205, 1099-1105
17. Seijffers, M. J., Miller, L. L., and Segal, H. L. (1964) Biochemistry 3, 1203-1209

18. Turner, M. D., Mangla, J. C., Samloff, I. M., Miller, L. L., and Segal, H. L. (1970) Biochem. J. 116, 397-404
19. Antonini, J., and Ribadeau Dumas, B. (1971) Biochimie 53, 321-329
20. Tang, J. (1970) in Methods in Enzymology (Colowick, S. P., and Kaplan, N. O. eds) 19, 406-421
21. Hanley, W. B., Boyer, S. H., and Naughton, M. A. (1966) Nature (London) 209, 996-1002
22. Samloff, I. M. (1969) Gastroenterology 57, 659-669
23. Etherington, D. J., and Taylor, W. H. (1969) Biochem. J. 113, 663-668
24. Mangla, J. C., Guarasci, G., and Turner, M. D. (1973) Am. J. Dig. Dis. New Ser. 18, 857-864
25. Shaw, B., and Wright, C. L. (1976) Digestion 14, 142-152
26. Samloff, I. M. (1971) Gastroenterology 61, 185-188
27. Samloff, I. M., and Liebman, W. M. (1973) Gastroenterology 65, 36-42
28. Zöller, M., Matzku, S., and Rapp, W. (1976) Biochim. Biophys. Acta 427, 708-718
29. Hirsch-Marie, H., Loisillier, F., Touboul, J. P., and Burtin, P. (1976) Lab. Invest. 34, 623-632
30. Rothe, G. A. L., Axelsen, N. H., Jøhnk, P., and Foltmann, B. (1976) J. Dairy Res. 43, 85-95
31. Kotts, C., and Jenness, R. (1976) J. Dairy Sci. 59, 1398-1400
32. Henschel, M. J. (1973) Brit. J. Nutr. 30, 285-296
33. Alais, C., Dutheil, H., and Bosc, J. (1962) Proc. XVI Int. Dairy Congr. Vol.B, 643-654
34. Lee, D., and Ryle, A. P. (1967) Biochem. J. 104, 735-741
35. Meitner, P. A., and Kassell, B. (1971) Biochem. J. 121, 249-256
36. Rajagopalan, T. G., Moore, S., and Stein, W. H. (1966) J. Biol. Chem. 241, 4940-4950
37. Foltmann, B. (1964) C. R. Trav. Lab. Carlsberg 34, 319-325
37a. Kageyama, T., and Takahashi, K. (1977) Biochem. Biophys. Res. Commun. 74, 789-795
38. Stepanov, V. M., Timokhina, E. A., Baratova, L. A., Belyanova, L. P., Korzhenko, V. P., and Zhinkova, T. G. (1971) Biochem. Biophys. Res. Commun. 45, 1482-1487
39. Inagami, T., Misono, K., and Michelakis, A. M. (1974) Biochem. Biophys. Res. Commun. 56, 503-509
40. McKown, M. M., Workman, R. J., and Gregerman, R. I. (1974) J. Biol. Chem. 249, 7770-7774
41. Knight, C. G., and Barrett, A. J. (1976) Biochem. J. 155, 117-125
42. Marks, N., Grynbaum, A., and Lajtha, A. (1973) Science 181, 949-951
43. Smith, G. D., Murray, M. A., Nichol, L. W., and Trikojus, V. M. (1969) Biochim. Biophys. Acta 171, 288-298
44. Ruenwongsa, P., and Chulavatnatol, M. (1975) J. Biol. Chem. 250, 7574-7578
45. Sardinas, J. L. (1972) Advan. Appl. Microbiol. 15, 39-73
46. Leckie, B. (1973) Clin. Sci. 44, 301-304

47. Ong, E. B., and Perlmann, G. E. (1968) J. Biol. Chem. 243,
    6104-6109
48. Pedersen, V. B., and Foltmann, B. (1973) FEBS Lett. 35, 255-256
49. Sepulveda, P., Marciniszyn, J., Liu, D., and Tang, J. (1975)
    J. Biol. Chem. 250, 5082-5088
50. Morávek, L., and Kostka, V. (1974) FEBS Lett. 43, 207-211
51. Revina, L. P., Yakhitova, E. A., Pugacheva, I. B., Lapuk, Y. I.,
    and Stepanov, V. M. (1972) Biokhimiya 37, 1074-1081
52. Stepanov, V. M., Baratova, L. A., Pugacheva, I. B., Belyanova,
    L. P., Revina, L. P., and Timokhina, E. A. (1973) Biochem.
    Biophys. Res. Commun. 54, 1164-1170
53. Sepulveda, P., Jackson, K. W., and Tang, J. (1975) Biochem.
    Biophys. Res. Commun. 63, 1106-1112
54. Huang, W.-Y., and Tang, J. (1970) J. Biol. Chem. 245, 2189-2193
55. Harboe, M. K., Andersen, P. M., Foltmann, B., Kay, J., and
    Kassell, B. (1974) J. Biol. Chem. 249, 4487-4494
56. Harboe, M. K., and Foltmann, B. (1975) FEBS Lett. 60, 133-136
57. Meitner, P. A. (1971) Biochem. J. 124, 673-676
58. Rasmussen, K. T., and Foltmann, B. (1971) Acta Chem. Scand. 25,
    3873-3874
59. Klemm, P., Poulsen, F., Harboe, M. K., and Foltmann, B. (1976)
    Acta Chem. Scand. Ser B,(in press)
60. Pedersen, V. B., and Foltmann, B. (1975) Eur. J. Biochem. 55,
    95-103
61. Pedersen, V. B., and Foltmann, B. (1973) FEBS Lett. 35, 250-254
62. Cunningham, A., Wang, H.-M., Jones, S. R., Kurosky, A., Rao, L.,
    Harris, C. I., Rhee, S. H., and Hofmann, T. (1976) Can. J.
    Biochem. 54, 902-914

X-RAY CRYSTALLOGRAPHIC STUDIES OF PEPSIN

N. S. Andreeva, A. E. Gustchina, A. A. Fedorov,
N. E. Shutzkever, and T. V. Volnova

Institute of Molecular Biology, Academy of Sciences
U.S.S.R., Moscow

Pepsin was the first object used to obtain x-ray photographs of protein crystals (1) and the second protein to be crystallized (2). J. Northrop developed the method of porcine pepsin crystallization in this well known investigation (2). Two of the crystal forms of pepsin that have been described, the first hexagonal form has been studied by J. Bernal and D. Crowfoot (1), and by M. Perutz (3). The unit cell of hexagonal pepsin crystals was too large, however, for studies of it to continue at the initial stage of the development of protein crystallography. The second form of the crystal was more suitable for x-ray crystallographic studies.

DATA COLLECTION

The second crystals described as thin needles (2), could be obtained from water-ethanol solutions at pH 2, the optimum of pepsin activity. This was one of the reasons we initiated x-ray crystallographic studies of this crystal form. Using a slightly modified method of pepsin crystallization, we could obtain thin plates of pepsin crystals approximately 1 x 0.3 x 0.1 mm. The crystals have the symmetry of the space group $P2_1$ with one molecule per asymmetric unit; cell dimensions are a = 54.7 A, b = 36.3 A, c = 73.5 A and $\beta = 103^{\circ}50'$. The diffractional field of these crystals has the minimum Bragg spacing of about 2.5 A. The heavy atom derivatives used in our analysis were prepared by soaking the native enzyme crystals in 20% ethanol-water solutions, pH 2, containing 0.001 M of various heavy atom compounds. The specific feature of this crystal was the binding of heavy atoms at two common sites for all the derivatives tested. Therefore, we could use only two derivatives: platinum nitrate and mercury iodide.

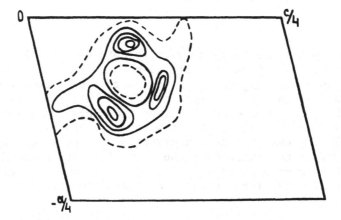

Figure 1. Double-difference Fourier synthesis for the
derivative HgI$_3$, after subtraction of the density at the
position of mercury.

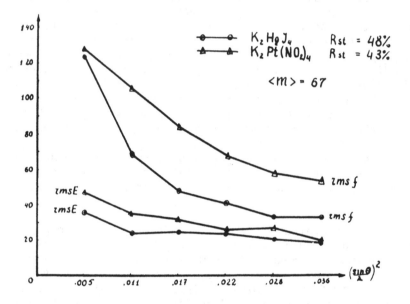

Figure 2. Reliability factors of the phase determinations.

The solution of the phase problem was based on the presence of the third strong site for platinum compound and on the special arrangement of iodine atoms in mercury iodide ion, which in this special case was trigonal. The contribution of this derivative into crystal scattering could be observed up to 3 A.

Data collection were performed by the conventional precession method. x-Ray photographs were obtained with the use of a Rigaku-Denky generator operated at 40 kV, 80 ma. The optical densities recorded on the packs of G-Ilford film were measured with a Joyce and Lobl microdensitometer and an Optronics scanner. Only one of these two methods has been constantly used for intensity measurements of the same reciprocal lattice plane recorded on x-ray photographs of the native compound and all the derivatives. The mean error in intensities on scaling symmetrically related spots - R_{sym} varied within the limits 3-7%, while film scaling error for each R_{scale} - 2-5%. The offline version of scanner has been used in this study; magnetic tape records were simultaneously processed for two films of each pack by means of the program kindly made available by R. Huber and P. Schwager and subsequently modified in our laboratory. The intensities were measured for approximately 6,500 Friedel pairs of reflections in a region corresponding to 2.7 A resolution.

Derivatives data were scaled to those of the native compound separately for each reciprocal lattice plane recorded on x-ray photographs. These data were corrected by the fall-off factor differences.

The determination of heavy atom parameters in pepsin crystals has been described in earlier publications (5). These parameters were refined by means of the H. Muirhead program (6) with the calculation residual Fourier maps after each refinement procedure. Reliability factors improved significantly after localization of all iodine atoms in mercury iodide ions. RMS-lack of triangle closure error fell from 47 to 24; as a consequence, the standard reliability factor was improved from 0.56 to 0.48 for this derivative (Figs. 1 and 2). "Best" MIR phases (7) were calculated, including the contribution from the measurements of anomalous scattering of the derivatives. Fourier synthesis was calculated in an arbitrary scale, and electron density distribution was sampled at about 1 A intervals along all three axes.

STRUCTURAL INTERPRETATION

There was enough contrast in this synthesis to see the segments of the polypeptide chain. Molecular boundaries were generally well defined; however, there were several points of close intermolecular contacts. It was also possible to detect protuberances of electron density spaced at intervals of about 4 A along many segments.

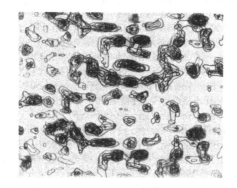

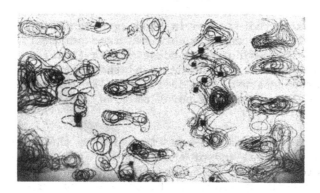

Figure 3. View of several sections of the Fourier synthesis:
(a,top) sections of 3.5 A resolution map containing NH_2-terminal
β-strands and the cleft; (b,bottom) sheets of 2.7 A resolution
map, mounted in Richard's box, containing residues 225 to 235-
α-helix near the surface of the molecule.

Therefore, we hoped to interpret this synthesis using the data on
the primary structure of porcine pepsin (8), despite many close
contacts between chain segments. Two regions of the synthesis are
presented in Figure 3.

The first attempt to interpret this synthesis was based on the
hypothesis formulated during low resolution work (5) concerning the
positions of heavy atom binding sites. Three heavy atom binding
sites of pepsin were located in a small region of the electron
density map near a helical-type segment. Pepsin contains four
basic amino acid residues; three were found at the small carboxyl-
terminal polypeptide (8,9). One could suppose that three binding
sites correspond to three basic groups of carboxyl-terminal region,
as only negatively charged ions were used for the isomorphous replacement. In this case the arrangement of the eight terminal
residues could correspond to the position of the helical segment.
The molecular boundaries chosen during low resolution work outlined
the surface of a molecule in this region in such a way that the
carboxyl-terminal segment formed a wall of a predominant depression.
Therefore, it was possible to think that the carboxyl-terminal segment was near the active site of the enzyme. Several chemical data
obtained after our studies were in good agreement with this hypothesis. The most important was obtained in experiments on the modification of Arg-316 residue of porcine pepsin (10,11). Additionally,
L. M. Ginodman and V. G. Grigoriev found that heavy atom compounds
used for the preparation of the derivatives inhibited the activity
of the enzyme (12). From this point of view, we tried to interpret
the synthesis.

The interpretation of the map started from the helical segment
at the carboxyl-terminal end of the molecule; then the chain was
traced in accordance with the approximate position of bulky side
chains. The preliminary model of the enzyme structure was built (13).
This interpretation met with difficulties, however, when the work
on model building of side chains with the use of mirror comparator
was developed. Another difficulty with this interpretation was the
presence of the large internal cavity in the molecule, which could
not be accessible for high molecular weight compounds.

The map was therefore reinterpreted for the revised positions
of the molecular boundary near the helical segment. The molecular
boundaries were also slightly changed in two other places where the
separation of molecules was also not well defined on the map. After
this modification, the entrance into the cavity became wider and the
two-domain structure of the enzyme more prominent. The dimensions
and the shape of the molecule became very similar to those of microbial acid protease (14).

Pepsin molecules are located between screw axes in the monoclinic unit cell. They are elongated along the x-direction.

Each molecule occupies almost completely the whole period of 54.7 A.
Intermolecular contacts are very close along this direction. In
another direction, a molecule has dimension slightly less than y-axis
(36 A); the third dimension corresponds approximately to one half of
z-axis. Only a very small part of each molecule is projecting beyond the screw axes. The molecule consists of two lobes, separated
by a uniquely shaped cleft. Symmetrically related molecules occupy
the entrance into the cleft of each other.

One lobe has a ball-shaped region of approximately 18 A diame-
ter. The main structural components of this lobe are strands of
the extended segments forming sheet structures of two and three
strands in various places. This region contains a loop which was
interpreted as the 45-50 cystine loop of porcine pepsin (8). The
ball-shaped region is connected to the body of a molecule through
segments extending to the active site. Beyond this region one can
see a projecting segment consisting of the helical-type turn and
the extended segment found in the residues 134-148 segment of microbial enzymes (14). (The chain tracing in this region performed
during preliminary interpretation is identical). The amino-terminal
end is located in this lobe and is far from the active site, same
as was found for the preliminary interpretation.

The feature of the other lobe is a helical segment with the
axis parallel to the monoclinic axis of the unit cell. The helix
is quite well-defined, but is rather short and can accommodate only
nine residues. Not as far from the helix there is a strong bridge
of density between strands forming a loop; this can be interpreted
as disulfide bridge Cys-250 and Cys-283. In our synthesis the
second lobe is presented by several high density strands, but there
is the empty region of a special shape corresponding to the space
near the axis of the barrel.

The active-site cleft of porcine pepsin has the wide entrance
(about 12 A), which decreases by one step due to the special geometry of the connection of the ball-shaped region to the main body of
a molecule. Electron density distribution in the active site of
porcine pepsin can be interpreted as the approach of the two loops
containing the active aspartic acids. There is a continuous bridge
of the electron density between two aspartic acids in our map, which
can be the consequence of their close contacts. In accordance with
this interpretation, the carboxyl-terminal end cannot approach the
active site. Though the close arrangement of several side chains
is visible in the active-site region, the accurate description of
this arrangement should be possible after the model building of the
active site, including side chains.

The data for the complexes of pepsin with several inhibitors
are being collected. The inhibitors used in this work are: diiodo-
tyrosyl-diiodotyrosine ethyl ester, phenylalanyl-diiodotyrosine

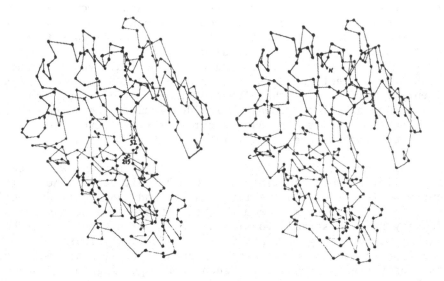

Figure 4.   Stereo-view of the conformation of pepsin chain.

ethyl ester and 1,2-epoxy-3-(p-iodophenoxy)propane. Difference
Fourier projections were calculated for all of them. To check the
reliability of various peaks on the different maps, we calculated
cross-difference Fourier synthesis for the derivatives by the use
of phases found from positions of iodine atoms of the inhibitors.
For the iodine atom of the inhibitor 1,2-epoxy-3-(p-iodophenoxy)
propane, two sites were found. The position of one site in pro-
jection corresponds to the cleft near the ball-shaped region.

In Figure 4, chain tracing for porcine pepsin is presented.
Although some details of this tracing can be improved, we believe
that it represents correctly the main structural features of the
enzyme molecule. Except for small deviations, there is a close
resemblance between the structure of pepsin and other acid proteinases, reflecting the homology of the primary structure of
these enzymes (15).

REFERENCES

1. Bernal, J. D., and Crowfoot, D. (1934) Nature 133, 794-795
2. Northrop, J. H. (1930) J. Gen. Physiol. 13, 739-766
3. Perutz, M. F. (1949) Research 2, 52-61
4. Bakulina, V. M., Borisov, V. V., Melik-Adamjan, V. R., Shutzke-
    ver, N. E., and Andreeva, N. S. (1968) Kristallogr. 13, 44-48
5. Andreeva, N. S., Borisov, V. V., Melik-Adamjan, V. R., Raiz,
    V. S., Trofimova, L. N., and Shutzkever, N. E. (1971) Mol.
    Biophys. 5, 908-916
6. Muirhead, H., Cox, J., Mazzarella, L., and Perutz, M. F. (1967)
    J. Mol. Biol. 28, 117-156
7. Blow, D. M., and Crick, F. H. C. (1959) Acta Crystallogr. 12,
    794-802
8. Tang, J., Sepulveda, P., Marciniszyn, J., Jr., Chen, K. C. S.,
    Huang, W.-Y., Tao, N., Liu, D., and Lanier, J. P. (1973) Proc.
    Natl. Acad. Sci. U.S.A. 70, 3437-3439
9. Stepanov, V. M., and Vaganova, T. I. (1968) Biochem. Biophys.
    Res. Commun. 31, 825-830
10. Huang, W.-Y., and Tang, J. (1972) J. Biol. Chem. 247, 2704-2710
11. Kitson, T. M., and Knowles, I. R. (1971) FEBS Lett. 16, 337-338
12. Andreeva, N. S., Ginodman, L. M., Borisov, V. V., Melik-Adamjan,
    V. R., Shutzkever, N. E., and Grigoriev, V. G. (1970) Abstr. Int.
    Biochem. Cong., Switaerland
13. Andreeva, N. S., Fedorov, A. A., Gustchina, A. E., Shutzkever,
    N. E., Riskulov, R. R., and Volnova, T. V. (1976) Dok. Akad.
    Nauk USSR, 228, 480-483
14. Davies, D. R., and Blundell, T. Personal Communication, unpub-
    lished data (See also Chapters 3 and 4 in this volume)
15. Foltmann, B., and Pedersen, V. B. (Chapter 1 in this volume)

THE CRYSTAL STRUCTURE OF AN ACID PROTEASE

FROM RHIZOPUS CHINENSIS AT 2.5 A RESOLUTION

E. Subramanian, M. Liu, I. D. A. Swan[*] and D. R. Davies

Laboratory of Molecular Biology, National Institute of
Arthritis, Metabolism and Digestive Diseases, NIH
Bethesda, Maryland 20014

The acid-protease from Rhizopus chinensis was first isolated by
Fukumoto, Tsuru, and Yamamoto (1). Its optimum pH for catalytic
activity was shown to be between 2.9 and 3.3 (1). Although its
complete sequence has not yet been determined, some limited sequence
data are available - particularly that of the 39 amino-terminal
residues (2,3) and of the residues in the immediate vicinity of the
catalytically active aspartic acid residues (4,5). These data show
that this enzyme has substantial sequence homology with porcine pepsin. Investigations of the kinetics of catalysis (6,7) have led
to proposals of an extended subsite specificity.

Conditions for obtaining large crystals suitable for x-ray
diffraction analysis have been described previously (8). In a preliminary report on a 5.5 A resolution structure analysis (9), the
molecule was described as bilobal with an extensive cleft region.
In addition, a preliminary report has been made of the three-dimensional structure of the enzyme as derived from a 3.0 A electron
density map, together with a comparison of this structure with the
structure of the acid protease from Endothia parasitica (10).

In this paper we discuss the results obtained from an examination of the 2.5 A resolution electron density map, and present some
of the data obtained from binding pepstatin (an acid-protease-speci-
fic inhibitor) to the enzyme in the crystals.

---

*Present Address: Department of Chemistry, University of Glasgow,
Scotland.

Table I
Summary of Heavy Atom Refinement Statistics

Resolution		6.79	4.71	3.97	3.55	3.26	3.05	2.89	2.75	2.64	2.54	Total	R*_C	R†_K
Number of reflections	N	1424	1336	1212	1188	965	1009	955	823	769	70~	10.05		
Figure of merit	$\langle m \rangle$	.93	.87	.84	.77	.72	.70	.73	.68	.65	.64	.77		
Native amplitude	$\langle F_P \rangle$	20.1	24.4	23.2	20.3	19.1	16.5	13.7	12.6	11.6	10.8	18.2		
Pb acetate: (3Å)	ε	2.22	2.51	2.54	2.58	2.69	2.67	2.31	2.13			2.50	.53	.09
	F_H	5.57	4.96	4.25	3.85	3.71	3.52	3.29	2.84			4.44		
Pb acetate: (3 to 2.5Å)	ε						1.41	1.37	1.34	1.26	1.25	1.32	.60	.08
	F_H						2.75	2.56	2.31	2.14	1.96	2.29		
Iodine: (3Å)	ε	2.51	2.81	2.90	2.76	2.64	2.57					2.71	.51	.09
	F_H	6.56	5.94	5.32	4.88	4.45	4.34					5.46		
Iodine: (3 to 2.5Å)	ε						2.45	2.21	1.99	1.94	1.92	2.06	.57	.12
	F_H						3.68	3.46	3.28	3.13	2.99	3.27		
Iodine + Pb: (3Å)	ε	4.00	4.21	4.29	3.99	3.75	3.67					4.03	.56	.14
	F_H	9.03	8.25	7.48	6.58	6.24	5.90					7.56		
UO2: (3Å)	ε	3.29	3.51	3.37	3.40	3.27	2.83					3.31	.55	.11
	F_H	7.10	6.04	5.35	4.99	4.50	4.54					5.72		
Merbromin: (4Å)	ε	1.90	2.44	2.63								2.25	.62	.07
	F_H	2.46	2.24	2.24								2.34		
K2IrCl6: (4.5Å)	ε	1.48	1.71									1.58	.50	.06
	F_H	2.97	1.81									2.57		
Billmann's mercurial: (5.5Å)	ε	2.32										2.32	.47	.07
	F_H	4.64										4.64		
Baker's mercurial: (5.5Å)	ε	3.73										3.73	.55	.13
	F_H	6.95										6.95		

*$R_C = \Sigma \; |||F_{PH}| - |F_P|| - |F_H|| / \Sigma \; ||F_{PH}| - |F_P||$, Σ over centric reflections only.

†$R_K = \Sigma \; ||F_{PH}| - |F_P + F_H|| / \Sigma \; |F_{PH}|$. ε is the lack of closure error of the phase triangle. F_P, F_{PH} and F_H are structure amplitudes for the parent enzyme, derivative enzyme and heavy-atom contribution, respectively. $\langle x \rangle$ represents averaging while x signifies root-mean-square averages.

MATERIALS AND METHODS

The Rhizopus chinensis acid protease was obtained from Miles/ Seikagaku. The crystallizing conditions reported by Swan (8) were modified by the addition of 50 mM cacodylate to ensure better buffering action. Large crystals 0.6 x 0.5 x 0.4 mm were obtained routinely by the use of ultra-filtration before the solutions were placed in the refrigerator for crystallization. The crystals are orthorhombic, $P2_12_12_1$, with a = 60.3, b = 60.7 and c = 107.0 A and one molecule per asymmetric unit (solvent content $\sim$ 56%).

Heavy atom derivatives were prepared by soaking the crystals in appropriate reagents for at least three days. Table I lists the various derivatives used to calculate the 2.5 A map.

Three-dimensional intensity data were collected on a Picker FACS-I computer-controlled diffractometer. Data from the crystals were rejected when there was more than 10% drop in the intensity of three standard reflections monitored every 2 hrs. Multiple observations were averaged and intensities for the various derivatives were scaled to the native data.

Except for the iodinated derivative, the major site of substitution of the heavy atoms was found from three-dimensional difference Patterson maps and verified by 'cross' difference-Fourier maps phased on other derivatives. Alternate cycles of phase calculation and refinement of heavy atom parameters were carried out until convergence was reached. Minor sites of substitution of the heavy atoms were located on Fourier maps in which the 'lack of closure' error for the derivative was used as coefficients (11). Table I summarizes the refinement statistics as well. The final figure-of-merit (12) for the 2.5 A data is 0.77.

Inhibitor studies were carried out by soaking the crystals in saturated solutions of the inhibitors for two weeks or longer. The crystals were then examined by x-ray photographic techniques for indications that the inhibitor had bound. Positive indications were followed by data collection on the diffractometer. Difference electron density maps were prepared in the usual way.

RESULTS AND DISCUSSION

Interpretation of the Map

The electron-dense regions were quite clearly resolved with respect to the surrounding solvent region, and permitted a fairly

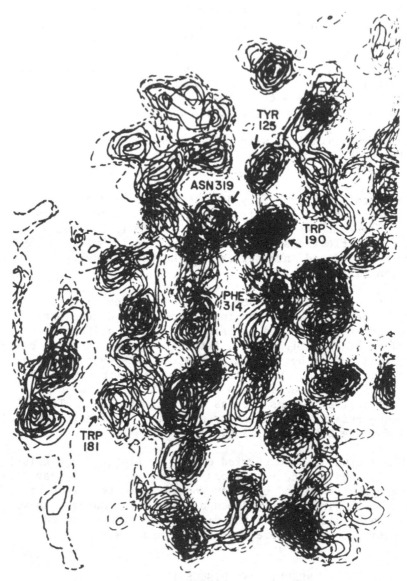

Figure 1:   Composite of a few sections from the 2.5 A electron
density map of the acid protease from Rhizopus chinensis,
which shows four anti-parallel β-strands.   Some of the prominent
side chains are labeled.   (The numbers used correspond to the
pepsin sequence).

unambiguous chain tracing from one terminus to the other. Most of the side chains were clearly visible, and aromatic side chains could be recognized as such; a representative section of density is shown in Figure 1. The side chain density for each of the 39 NH_2-terminal residues could be fitted well to the known sequence. More interestingly, however, assuming structural homology with pepsin on the basis of similarity in amino acid compositions, (in particular, by assuming that the number and location of the hydrophobic residues were the same), we were able to identify from the map most of the aromatic side chains which are conserved for most acid proteases (see Chapter 1).

The amino acid composition of the Rhizopus enzyme indicates that there are two S-S bridges in the molecule. We could clearly locate one of them at 46-50 (the numbering corresponds to the pepsin sequence). The second S-S bridge might be expected to correspond to either 206-210 or to 250-283 of pepsin. Residues 250 and 283 were at the right distance for an S-S bridge, and there was reasonable density in the map between these two residues. The polypeptide chain was linear in the region corresponding to 206-210 and could not accomodate an S-S bridge. It is interesting that in the sequence alignment of Foltmann, (see Chapter 1) there is a gap of 3 residues in penicillopepsin in the exact location of this disulfide loop of pepsin.

Description of the Structure

From the map, we were able to pick out 325 α-carbon positions (the amino acid composition suggests 324 residues). Figure 2 is a drawing of the α-carbon backbone for the Rhizopus enzyme, viewed along the crystallographic b-axis.

The molecule consists of two lobes of approximately the same size with a pronounced cleft between the lobes. These lobes are connected by a single piece of polypeptide chain, so that one lobe contains the NH_2-terminal part while the COOH-terminus is situated in the other lobe. The molecule has the approximate overall dimensions of 37 x 46 x 63 A. The length of the cleft between the lobes is difficult to define precisely, but is at least 25 A.

The two lobes of the molecule are made up essentially of antiparallel β-strands. There are only four short helical regions. Three of them lie on the surface of the molecule and away from the cleft, and one near the base of the cleft. The helices contain from 1½ to 2½ turns. A frequent feature is the presence of several anti-parallel β-loops - a stretch of the chain turning back and running antiparallel to itself before branching off to form a similar loop elsewhere. The two lobes have contacts over a large area giving

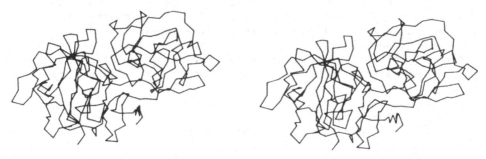

Figure 2:   Stereo drawing of the α-carbon chain tracing for the
Rhizopus enzyme, viewed along the crystallographic b-axis.
The NH_2-terminal end of the chain is located in the right
hand lobe.

rise to a broad waist. The COOH-terminal lobe consists of eleven extended β-strands, which together with one short helical region, are wrapped around what is probably a hydrophobic core.

The NH$_2$-terminal lobe consists almost entirely of extended segments of antiparallel β-strands. It comprises three sheets stacked on top of each other, each containing three strands, with the direction of the strands changing by approximately 90° in going from one sheet to another. This lobe includes a well-defined pocket, located at one side of the cleft. This pocket contains a site of iodination, and the presence of large aromatic side chains suggests a hydrophobic environment. This may be a specificity pocket. It is interesting that both pepsin (13) and acid protease from Mucor pusillus (14) lose their catalytic activity on iodination. The outer surface of this pocket consists of an extended β-loop around residue 80, forming a 'flap' over the hydrophobic interior. The density in the map for this loop is low, suggesting that it may be flexible.

The chain tracing locates the catalytically active Asp-35 at one end of the active site cleft. A residue that could correspond to the other catalytically active aspartic acid, namely Asp-215 (pepsin numbering), is located in close proximity to Asp-35. This region of our map is somewhat complicated by the fact that it contains the major binding site for most of our heavy atom derivatives, and is consequently difficult to interpret unambiguously.

Pepstatin Binding

Figure 3 is a 3.7 A map of the difference between the electron densities of the pepstatin-bound enzyme and the native enzyme, superimposed on the relevant sections of the electron density map for the native enzyme. It is clear that pepstatin binds in the cleft, and identifies this as the region of active site binding. The density on the difference map corresponding to the pepstatin molecule passes very close to the side chain density of Asp-35. Some of the pepstatin density also lies under the flap region referred to above. The general appearance of the map suggests that pepstatin binds in an extended conformation, covering a length of about 20 A. This is consistent with proposals from kinetic studies (7) of an extended binding site for acid-proteases. This result is preliminary, and we are in the process of extending the x-ray analysis of the pepstatin-enzyme complex to 2.5 A resolution.

Acknowledgment

We wish to thank Professor H. Umegawa for a generous gift of pepstatin.

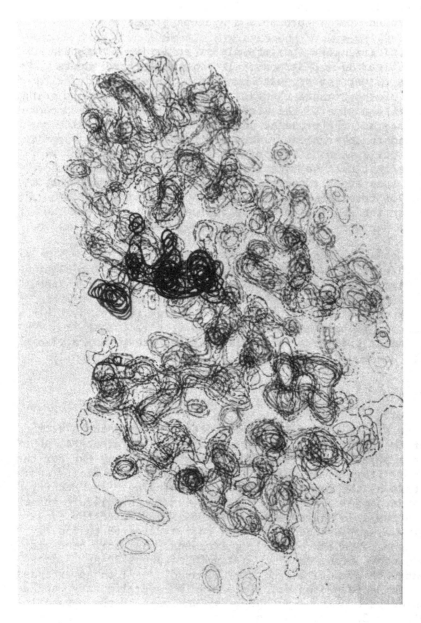

Figure 3: Pepstatin difference map (3.7 A resolution) superimposed on the relevant sections of the 2.5 A native map. The thick black lines represent the pepstatin contours.

SUMMARY

This paper contains a preliminary report of the crystal structure of the acid protease from Rhizopus chinensis at 2.5 A resolution. The molecule is bilobal with a large cleft between the lobes. Pepstatin binds in the cleft near the catalytically active Asp-35. The overall folding of the molecule consists primarily of antiparallel β-strands, there being only four small helices.

REFERENCES

1. Fukumoto, J., Tsuru, D., and Yamamoto, T. (1967) Agr. Biol. Chem. 31, 710-717
2. Gripon, J. C., Rhee, S. H., and Hofmann, T., Private Commun.
3. Sepulveda, P., Jackson, K. W., and Tang, J. (1975) Biochem. Biophys. Res. Commun. 63, 1106-1112
4. Graham, J. E. S., Sodek, J., and Hofmann, T. (1973) Can. J. Biochem. 51, 789-796
5. Chen, K. C. S., and Tang, J. (1972) J. Biol. Chem. 247, 2566-2574
6. Voynick, I. M., and Fruton, J. S. (1971) Proc. Natl. Acad. Sci. U.S.A. 68, 257-259
7. Sachdev, G. P., Brownstein, A. D., and Fruton, J. S. (1975) J. Biol. Chem. 250, 501-507
8. Swan, I. D. A. (1971) J. Mol. Biol. 60, 405-407
9. Subramanian, E., Swan, I. D. A., and Davies, David R. (1976) Biochem. Biophys. Res. Commun. 68, 875-880
10. Subramanian, E., Swan, I. D. A., Liu, Mamie, Davies, D. R., Jenkins, J. A., Tickle, I. J., and Blundell, T. L. (1977) Proc. Natl. Acad. Sci. U.S.A.(in press)
11. Subramanian, E., and Cohen, G. H. (1977) Unpublished
12. Blow, D. M., and Crick, F. H. C. (1959) Acta Crystallogr. 12, 794
13. Fruton, J. S. (1971) The Enzymes (P. D. Boyer, ed) Vol. 3, 3rd ed., p.119, Academic Press, New York
14. Arima, K., Yu, J., and Iwasaki, S. (1971) Methods Enzymol. 19, 446

X-RAY ANALYSIS AND CIRCULAR DICHROISM OF THE ACID

PROTEASE FROM ENDOTHIA PARASITICA AND CHYMOSIN

John Jenkins, Ian Tickle, Trevor Sewell,
Luciano Ungaretti*, Axel Wollmer[§], and Tom Blundell

Laboratory of Molecular Biology, Department of Crystallography
Birbeck College, London University, Malet Street, London WC1E   7HX

Laboratory of Biochemistry, School of Biological Sciences
Sussex University, Falmer, Brighton, Sussex

Central to the study of mechanism and specificity of the acid proteinases is a knowledge of the three-dimensional structures of enzymes with different physiological roles. The availability of acid proteinases with either extracellular or intracellular roles in vertebrates as well as similar enzymes from plants, protozoa, and fungi allows a wide range of comparative studies. In our laboratory we have undertaken the x-ray analysis of fungal enzymes, those from Endothia parasitica and Mucor pusillus, and some vertebrate enzymes including chymosin and chicken pepsin, and are beginning work with cathepsin D.

In this discussion we will be principally concerned with the crystal structure analysis of the acid proteinase from Endothia parasitica, an enzyme that is easily purified and crystallized (1,2). It rapidly clots milk at pH 5-6.5 and hydrolyzes casein and hemoglobin at maximum rates at pH 2-2.5 (3). Hydrolysis of the oxidized B-chain of insulin shows that this proteinase has a specificity for

---

*Present address: Centro di Studio per la Crystallografia Struttur-
ale del C. N. R., University of Pavia, Via A. Bassi 4, 27100, Pavia,
Italy.

[§]Present address: Abteilung Physiologische Chemie, der medizinischen
Fakultat, an der Rhein - Westf. Techn. Hochschule Aachen, 51 Aachen.

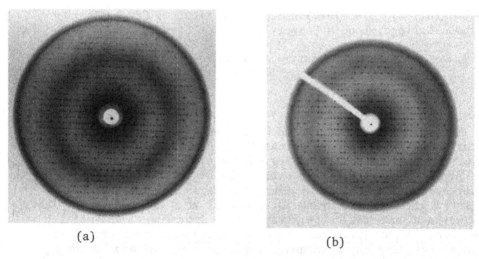

(a)   (b)

Figure 1.   x-Ray diffraction patterns (0kl zone) for (a) native and (b) K_2PtCl_4 derivative of the acid proteinase from _Endothia parasitica_.

the hydrophobic regions similar to that of pepsin (3). Peptide bonds
at B4Gln-B5 His, B11Leu-B12Val, B15Leu-B16Tyr, B16Tyr-B17Leu, B24Phe-
B25Phe and B25Phe-B26Tyr are hydrolyzed by both enzymes. The acid
proteinase from Endothia parasitica, however, hydrolyzes bonds at
B3Asn-B4Gln and B20Gly-B21Glu, not attacked by pepsin, but fails to
hydrolyze bonds involving B1Phe-B2Val, B13Glu-B14Ala, B14Ala-B15Leu
and B23Gly-B24Phe, hydrolyzed by pepsin. The preliminary x-ray
crystallographic study was made by Dr. P. C. Moews and Dr. C. W.
Bunn (4), and we inherited the problem on Dr. Bunn's retirement in
1972. We have reported a low resolution electron density map (5)
which showed the enzyme to be bilobal with a well-defined and extensive cleft. We now report a medium (3 A) resolution electron density
map which defines the general topography and allows a tentative
identification of active site residues assuming the enzyme homologous
with pepsin.

We also have crystals of chymosin (6), the acid proteinase from
Mucor pusillus (7), chicken pepsin (8), and chicken pepsinogen (8),
the first two of which are large and very suitable for x-ray analysis.
Dr. C. W. Bunn and his co-workers made preliminary x-ray studies of
chymosin but the method of isomorphous replacement was unsuccessful,
as simple heavy atom derivatives proved impossible to prepare. Both
we and Professor B. Foltmann and Dr. S. Larsen of Copenhagen have
continued x-ray studies on chymosin, but have also met difficulties.
More recently, we have began to use the structural information from
the Endothia parastica enzyme to solve the structure of chymosin by
using the method of molecular replacement.

Similar use of molecular replacement combined with refinement
at high resolution appears to be the most hopeful method of defining
the molecular structures of the enzymes from Mucor pusillus and
chicken pepsin.

X-RAY ANALYSIS OF THE ENDOTHIA PARASITICA ENZYME

The enzyme was extracted by the method of Hagemeyer et al. (1)
from "Sure Curd" kindly donated by Dr. J. L. Sardinas of Pfitzer
Chem. Co. Crystals were grown from 2.7 M ammonium sulfate plus 5-10%
acetone using batch techniques or dialysis at pH's between 4.5 and
6.3 in 0.1 M acetate or phosphate buffer (4). The crystals, nearly
rectangular plates of dimensions 1.2 x 0.5 x 0.1 mm, give x-ray
diffraction patterns (see Fig.1a) to at least 2 A resolution. The
space group is $P2_1$ with cell dimensions of a = 53.6, b = 74.05,
c = 45.7 A, β = 110°. If there is one molecule of the enzyme in the
asymmetric unit, the crystal solvent is 49% by volume (4). x-Ray
data was collected on a Hilger-Watts 4-circle diffractometer using
a program written by one of us (9). Two equivalent reflections were
measured for native data to 2.3 A resolution using seven crystals.

Three standards were measured every 51 reflections and crystals were
discarded if the intensity of any fell by 20% of its initial value.
Data have been collected from crystals of six heavy atom derivatives.
Friedel pairs were collected to allow estimates of the contribution
of anomalous scattering. Absorption corrections were calculated
using the empirical method of North et al. (10).

A survey of over 40 possible heavy atom reagents suggested that
the crystals were comparatively unreactive. In some cases reaction
was attempted at both pH 4.5 and 6.3. The heavy atom salts listed
in Table I gave reasonable heavy atom derivatives, although in most
cases the reaction conditions needed to be carefully controlled to
avoid loss of isomorphism. Estimates of F_H, the magnitude of the
contribution of the heavy atom to the structure factor, were calculated from isomorphous differences and anomalous differences to give
combination coefficients (11-15). The combination difference
Pattersons allowed the determination of the heavy atom sites; the
positions and thermal parameters were refined using least squares
refinement with F_{HLE} values as observed structure factor amplitudes
(see Ref.15).

The uranyl heavy atom parameters were then used to calculate
F_H and F_H''; two sets of phases for the protein were calculated - one
for each possible heavy atom enantiomorph - by the method of single
isomorphous replacement with the inclusion of anomalous scattering
data. The value of E'' was initially calculated from the r.m.s.
error in the least squares refinement for the centric zone, and E''
was made equal to E/3. The phases, α_D were used to compute difference Fouriers with coefficients {F_{PH} - F_P}exp.$i\alpha_p$ for the other derivatives. The phases calculated from uranyl positions gave a clear
indication of the heavy atom positions of other derivatives which
agreed well with those positions determined from the Patterson
functions. The enantiomorphic set of uranyl positions gave no clear
indication of the heavy atom positions. Thus, the correct enantio-
morphs and relative origins for derivatives were established. We
then carried out a series of "phase refinement" cycles (16,17).
The phases, α_p were initially calculated from sets of derivative
data excluding the derivatives refined; the "lack of closure" was
minimized for reflections with figure of merit, m>0.6 by least
squares refinement of positional and isotropic thermal parameters
and relative scale factor for the third heavy atom derivative. This
was repeated for each derivative and cycled until the refinement
approached convergence. In the latter cycles the positional parameters and occupancies were refined in zones of different mean
$\sin^2\theta/\lambda^2$ values. E and E'' values used in phase calculation (18,19)
were estimated by comparing observed and calculated F_{PH} and $F_{PH}(+)$ -
$F_{PH}(-)$ values using phases based on all derivatives. Finally, these
were included in a calculation of phases giving an average figure of
merit, $\bar{m}$ = 0.70. Values for the phasing power are shown in Table I.

Table I

Statistics for Endothia Parasitica Acid Proteases

Derivative	Number of sites	Range in $\sin^2\theta/\lambda^2$	$\bar{F}_H$	E	$\bar{F}''$	E''	$(\bar{F}_{PH} - \bar{F}_P)$
Uranyl acetate	5	.0000 – .0096	110	48	20	32	86
		.0096 – .0192	83	38	17	35	63
		.0192 – .0288	64	46	13	60	71
K_2PtCl_4	1	.0000 – .0096	77	47	–	–	72
		.0096 – .0192	64	48	–	–	62
Platinum $(en)_2Cl_2$	2	.0000 – .0096	65	47	–	–	67
		.0096 – .0192	49	40	–	–	52
		.0192 – .0288	36	58	–	–	59
K_2HgBr_4	2	.0000 – .0096	80	56	9	28	76
		.0096 – .0192	54	66	5	41	69
		.0192 – .0288	37	57	4	44	64
$KAu(CN)_2$	3	.0000 – .0096	93	72	–	–	97
		.0096 – .0192	76	72	–	–	79
KI_3	7	.0000 – .0096	108	68	–	–	97
		.0096 – .0192	94	85	–	–	115
		.0192 – .0288	85	89	–	–	120

	Range in $\sin^2\theta/\lambda^2$		
	.0000 – .0096	.0096 – .0192	.0192 – .0288
Statistical parameter $\bar{F}_P$	370	407	292
Figure of merit	.89	.74	.57

The parameters finally obtained by minimization of "lack of closure" were in good agreement with those calculated from F_{HLE} or $|F_{PH} \pm F_P|$ refinement in centric zones.

The best phases α_p and figures of merit were used to calculate a 5.2 A resolution electron density map (Fig.2), and a 3 A resolution map (Fig.3). This map was calculated in October 1975, using the uranyl acetate, Pt(ethylenediamine)Cl$_2$ and K_2HgBr_4 derivatives to 3 A resolution and data to 4.5 A on both K_2PtCl_4 and $KAu(CN)_2$ together with the data to 3.3 A from the iodine crystal. It is on this electron density map that most of the interpretation has been carried out. In the summer of 1976, a further electron density map at 2.7 A was calculated using data from the uranyl acetate derivative from 3.0 to 2.7 A and mercuric acetate derivative, and a new set of iodine derivative data to 3.5 A resolution. This is closely similar to the 3.0 A electron density in general, but showed sharpening and shifting of some peaks and an improved density at several difficult points thought to represent the main chain. Further improvement of the resolution and quality of the electron density map is limited by lack of isomorphism of the heavy atom derivatives. Extension of the uranyl acetate derivative data to 2.5 A and the iodine derivative to 3.0 A, and possibly including a K_2HgI_4 derivative, represents the limit of usefulness of the method of multiple isomorphous replacement on this crystal form.

THE MOLECULAR STRUCTURE OF THE ENDOTHIA PARASITICA PROTEINASE

The 5.2 A resolution electron density map in Figure 2 has regions of low density, which appear to correspond to disordered solvent and a volume of higher density which has an overall ellipsoidal shape with major axes 60 x 46 x 37 A. This corresponds to the enzyme molecule, the boundary of which is indicated in Figure 2. The heavy atom positions lie on the surface of the molecule which is consistent with their interactions with surface side chains.

The higher resolution electron density maps confirm the molecular boundary (20). There are two areas of close intermolecular contacts, between three molecules in the x, z plane and between the two molecules separated in y, but this does not lead to an ambiguity in interpretation. The electron density allows the polypeptide chain to be traced over more than 300 residues. Side chains are usually clearly defined although the map does not allow clear orientation of the carbonyl groups. The major difficulties in interpretation arise in places where the electron density runs between chains in sheet structures, and this is particularly true of the polypeptide in the cleft. The enzyme is organized into two lobes which are joined in a broad waist giving a long 25 A cleft indicated in Figure 4. Figure 5 is a stereo pair of the molecule taken from the same direction. The straight line represents virtual bonds between α-carbon positions.

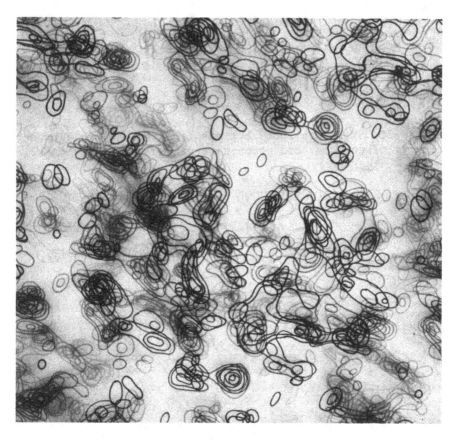

Figure 2. The 5.2 A electron density map of the acid proteinase viewed along the b̲ axis.

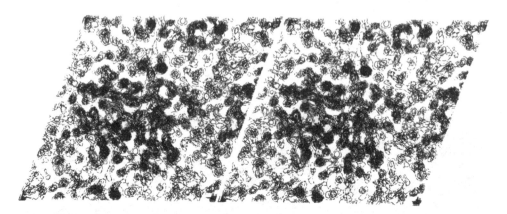

Figure 3. Stereo pair of sections from the 3.0 A electron density map of the acid proteinase viewed down the b axis.

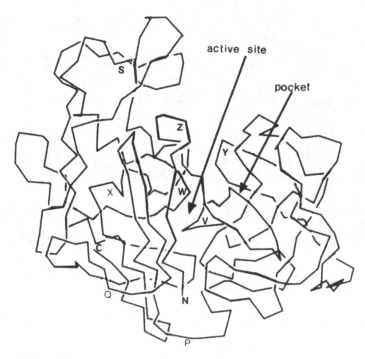

Figure 4. A schematic diagram of the structure of the acid proteinase, viewed down the b axis. The lines indicate the virtual bonds between alpha carbon positions. The letters are explained in the text.

The two chain termini are in strands 1 and 5 of a twisted sheet structure. The NH_2-terminus is inferred to be at N (Fig.4) from the disposition of side chains in the helices and by analogy with the pepsin sequence (vide infra). The first 140 residues of the chain fold to give the lobe on the right-hand side of Figures 4 and 5. The chain is led back into the sheet structure (P in Fig.4) at strand 3 of the sheet and out at strand 2 (Q in Fig.4). The carboxyl-terminal half of the polypeptide is folded to give the lobe on the left-hand side of Figure 4 and finally rejoins the sheet at strand 4 folding to place the COOH-terminus in strand 5 at C (Fig.4).

The NH_2-terminal lobe on the right-hand side of Figures 4 and 5 is predominantly sheet structure made from a series of antiparallel pairs of strands connected through tight bends. It is organized in three layers; the central layer is a sheet of five strands while the outer layers each comprise two loops of polypeptide with a chain direction at right angles to the polypeptide of the central sheet. As viewed in Figures 4 and 5, the upper and central layers are in close contact, but one of the lower loops is arranged to give a well-defined pocket leading from one side of the cleft. There are two small sections of helical structure on the surface of the lobe. The carboxyl-terminal lobe on the left-hand side of Figures 4 and 5 comprises a barrel of funnel of distorted sheet and two short sections of helix, one of three turns on the exterior of the molecule, and the other, which is rather shorter in the cleft region.

Contacts between the two lobes are very extensive. The strands in the β-sheet at the COOH-terminus contribute not only to the barrel, but also extend tangentially into the twisted sheet containing the NH_2-terminus. The fold in the NH2-terminal 20 residues extends the sheet structure into the NH_2-terminal lobe. The enzyme can therefore be described as an extended sheet folding into a sandwich structure in the NH_2-terminal lobe and the barrel in the COOH-terminal lobe. There are further interactions between the bottom of the β-barrel and the bottom of the sandwich, and many large, almost certainly aromatic groups, are thrown into the regions between and within the lobes and presumably give a hydrophobic core.

CIRCULAR DICHROISM OF THE ENDOTHIA PARASITICA ENZYME

The circular dichroism spectrum of the acid proteinase from Endothia parasitica was measured on a Cary61 CD spectrometer at $27^O C$, at a protein concentration of 0.232 mg/ml in 0.05 M phosphate buffer at pH 4.48. The path length was 10 mm for the near u.v. and 1 mm for the far u.v. The far u.v. circular dichroism is shown in Figure 6b. The spectrum resembles that of pepsin (21) and pepsinogen (22). The spectrum between 195 and 240 nm has been fitted using three sets of standard curves (23-25) with a least squares program; (written by

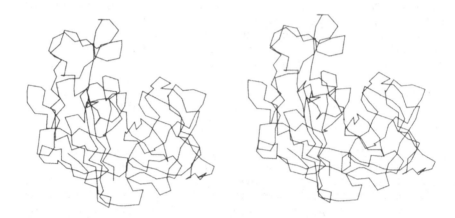

Figure 5. A stereo pair of the acid proteinase viewed down
the b̲ axis.

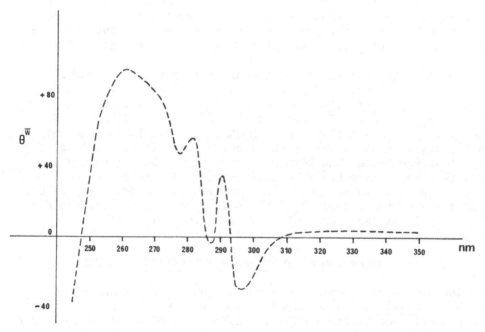

Figure 6a. The circular dichroism of the acid proteinase at
27ºC at a concentration of 0.232 mg/ml in 0.05 M phosphate
buffer at a pH of 4.48 (a) near ultraviolet.

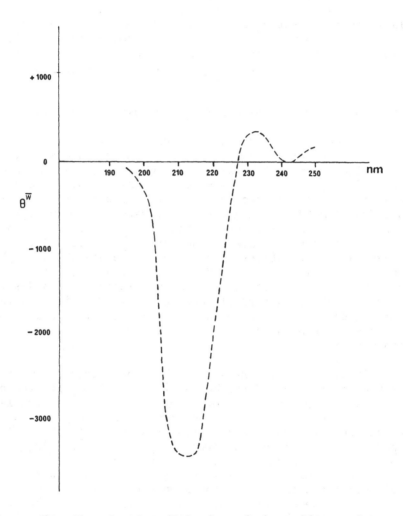

Figure 6b. The circular dichroism of the acid proteinase at 27°C at a concentration of 0.232 mg/ml in 0.05 M phosphate buffer at a pH of 4.48 (b) far ultraviolet spectra.

I. J. T.) the results are given in Table II. Although these techniques must be treated with caution, it is of interest that the low
percentage of helix and high proportion of β-sheet structure are consistent with the structure described from the x-ray analysis. The
near u.v. circular dichroism spectrum (see Fig.6a) shows a complex
structure strikingly similar to that found for penicillopepsin (26)
particularly in the position of the negative extrema. This may be
due to either trytophan or tyrosine as suggested by Wang et al. (26)
and suggests that the relative orientations and distances of the
residues involved have been conserved between the two enzymes.

MOLECULAR REPLACEMENT AND THE STRUCTURE OF CHYMOSIN

Chymosin (rennin, EC 3.4.4.3), the enzyme which degrades proteins in the stomach of the calf, is an acid proteinase which, on the
basis of an almost complete sequence (see Chapter I) is homologous
with pepsin (27,28). The x-ray studies of Bunn and his coworkers (6)
showed that large orthorhombic crystals are produced in the pH range
5.0 to 6.2 from 2 M NaCl solutions. The crystals have spacegroup
either I $2_1 2_1 2_1$ or I 222 with cell dimensions $\underline{a}$ = 79.7, $\underline{b}$ = 113.8
and $\underline{c}$ = 72.8 A and one molecule in the asymmetric unit. Attempts to
prepare heavy atom derivatives for use in the method of isomorphous
replacement proved difficult in 2 M NaCl solutions, and so Bunn
crosslinked the crystals with glutaraldehyde which allowed desalting.
In this way, the chymosin crystals became highly reactive and many
heavy atom reagents gave rise to intensity changes in the diffraction
pattern. However, conventional techniques were unable to locate the
heavy atom positions.

The difficulties in the x-ray analysis could arise from either
a complex binding pattern of the heavy atoms combined with the ambiguity in spacegroup, from lack of isomorphism of the derivatives,
or from some inherent disorder in the crystals possibly related to
their unusual sector structure reported by Bunn et al. (6). Although
it is impossible to rule out the latter explanation, the degree of
heavy atom substitution can be minimized by controlling the extent
of reaction. Unfortunately, our attempts to do this by working at
higher salt concentrations have proven unsuccessful, and we are now
investigating the use of the structural information from the acid
proteinase in solving the crystal structure of chymosin by molecular
replacement. This technique involves comparison of the Patterson
function of a crystal of unknown structure with that of a homologous
molecule, in order to determine their relative orientations and
positions in the crystal lattice. The first step is to identify and
isolate the electron density corresponding to a single molecule in
the known crystal structure, viz. Endothia parasitica proteinase.
This was done with sufficient accuracy by delineating the molecular
boundary by means of polygons drawn on the electron density map sections.

Table II

% Secondary Structure Predicted from Circular Dichroism (Far U.V.)

% HELIX	% BETA	% RANDOM	R VALUE
8.1 (0.7)	58.9 (1.4)	33.0 (0.8)	0.166 (1)
8.2 (5.3)	34.0 (7.3)	57.7 (3.3)	0.646 (2)
3.0 (4.2)	52.9 (4.6)	44.1 (1.5)	0.574 (3)

	MODEL	REFERENCE
(1)	10 proteins	Chen, Yang, and Chau
(2)	3 proteins	Saxena and Wetlaufer
(3)	Polylysine	Greenfield and Fasman

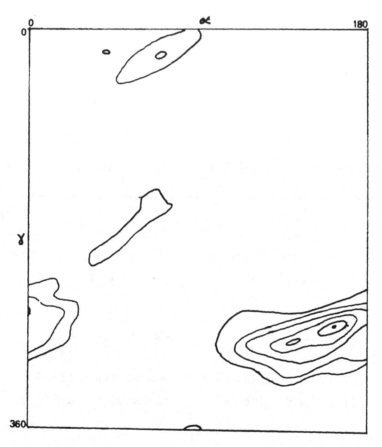

Figure 7.   Rotation function (5 A resolution) for chymosin versus <u>Endothia</u> <u>parasitica</u> proteinase, section β = 55° Maxima are at α = 161°, β = 56°, γ = 263° and α = 140°, β = 53°, γ = 277°.   Contour levels from 35 to 75 in steps of 10 (arbitrary units).

The diffraction intensities calculated for the molecular electron density were correlated with those observed for the chymosin crystal at 5 A resolution by means of a Fast Rotation Function program (29). The rotation function obtained indicated that two orientations are consistent with the data (see Fig.7); these are related by a rotation of 17° about an axis along the length of the cleft.

The significance of the two peaks in the rotation function is at present unclear. They may arise from disorder of the molecules in the chymosin crystals or more probably from a different relative position of the two lobes of the enzymes. These possibilities may be distinguished by carrying out the rotation function with the two lobes separately; this experiment is now underway in our laboratory. The results of the rotation function are strong evidence for a homology between chymosin and the acid proteinase from Endothia parasitica even though there may be some conformational change or disorder. Having fixed the orientation of the chymosin molecule as a whole, or of the two lobes separately, a translation function should allow the correct position in the unit cell to be determined. Assuming the chymosin crystals are not disordered between two orientations, it should be possible to calculate then an electron density map of the chymosin structure.

STRUCTURAL HOMOLOGY WITH OTHER ACID PROTEASES AND THE NATURE OF THE ACTIVE SITE OF THE ENDOTHIA PARASITICA ENZYME

The evidence from the rotation function that the enzyme from Endothia parasitica is homologous with chymosin and therefore with pepsin is supported by two further arguments. First, a comparison of the three-dimensional structure with that of the enzyme from Rhizopus chinensis shows a striking similarity (see Ref.20 and Chapter 3 in this book); as the first 40 residues of the Rhizopus enzyme are clearly homologous with pepsin, a similar homology for the enzyme from Endothia parasitica is implied. Secondly, the polypeptide chain is of similar length to that of pepsin and visits the centre of the cleft in a way which brings two residues equivalent to Asp-32 and Asp-215 into correlations between the structure of the Endothia parasitica enzyme and pepsin.

The NH_2-terminus (N in Fig.4) is quite distant from the centre of the cleft, but the first 20 residues form a loop with a β-bend at H which is close to the cleft. A further β-bend brings the chain back into the cleft and the active site aspartate equivalent to pepsin Asp-32 is probably at V. The chain reverses direction to form one side of the pocket (see Fig.4). A well-defined, large side chain in the pocket is probably a tryptophan equivalent to Trp-39.

A loop is formed to bring residues 45 and 50 into a position
where a disulfide might be formed in a homologous pepsin structure.
Residues 70-80 form a further loop on the other side of the pocket;
a large aromatic group which is iodinated in our KI_3 heavy atom
derivative may be analogous to Tyr-75 of pepsin. The arrangement
of large aromatic groups in this pocket may be related to the unusual circular dichroism spectrum in the near ultraviolet (vide
supra). Residues in the region of the 120 are also in the cleft
and the chain runs parallel to the strand containing "Asp-32". The
chain occupies a surface position in the region of residues 130-140
including a short helix (shorter than in Rhizopus) but comes close
to the cleft at residues around 150-160 in the sheet containing the
NH_2- and COOH-termini.

The first 30 residues of the COOH-terminal lobe are folded to
give three approximately antiparallel strands with two β-bends
before the chain visits the cleft in a loop at W, which appears to
be the best candidate for the second active site aspartate analagous
to Asp-215 of pepsin. If this is so, there must be a deletion of
some residues in this part of the chain, including the pepsin disulfide loop 206-210 for which there is no evidence in the enzyme
from Endothia parasitica. There is a sharp bend (pepsin residue
Gly-223) followed by a well-defined surface helix corresponding to
pepsin residues 225-238. Residues corresponding to pepsin 250 and
280 are bridged at S (Fig.4); and this appears to be the single
disulfide in the enzyme from Endothia parasitica. The chain at
residue 300 lies parallel to that containing the active site aspartate (Asp-215). There is a turn of helix around residue 305 so that
residue 308 is buried at one side of the cleft at X. As this is
either lysine or arginine in all other acid proteinases, it is
probably also basic in the enzyme from Endothia parasitica and its
position close to the cleft makes it a candidate for involvement in
the catalytic mechanism. Residue 316 is distant from the centre of
the cleft and is probably not important. The COOH-terminus is at C.

The two lobes define a deep and extensive cleft with a well-
defined and probably hydrophobic pocket. The cleft appears to be
wider at Y than at Z. At the centre of the cleft the residues
corresponding to Asp-32 and Asp-215 are in close proximity. A peptide chain of a substrate could be accommodated on either side of
these active site groups although a slight relative displacement
may be required one side near Z (Fig.4); the extended chain might
interact through a β-sheet structure with the chain of residues
close to "Asp-32". The position of the NH_2-terminus distant from
the cleft means that autolysis of zymogens must occur at the NH_2-
terminus either by an intermolecular reaction, or, alternatively,
the NH_2-terminal residues must have a different conformation in the
zymogen if the reaction is unimolecular. In view of the integral
nature of the NH_2-terminal strand to the overall structure, the
latter seems unlikely.

More probably, the extra residues in the zymogen are folded around the waist of the enzyme. If they were folded antiparallel to residues 1 to 10 of the enzyme, it is possible that in pepsin the acid groups (Asp-3, Glu-4, Glu-7) interact with basic groups of the zymogen.

Further interpretation of the structure and its relation to the catalytic activity of this acid protease depends critically on the determination of the primary structure which is now being undertaken by Professor B. Foltmann and Dr. V. Pedersen. It also depends on the co-crystallization of the enzyme with active site-directed inhibitors; unfortunately, we have been able to bind neither aromatic eposides nor pepstatin to the crystalline enzyme.

ACKNOWLEDGMENTS

We would like to thank Dr. C. Bunn and Dr. P. Moews for the initial gift of crystals and for making available their preliminary x-ray data. We would also like to thank Dr. J. L. Sardinas of Pfitzer Chemical Company for the kind gift of "Sure Curd". We are grateful to the Science Research Council for support of this project.

SUMMARY

The structure of an acid proteinase from Endothia parasitica has been solved by x-ray diffraction using multiple isomorphous replacement. A 3 A resolution map was interpreted in terms of a bilobal structure with a long 25 A cleft. The secondary structure is mostly distorted β-sheet. The circular dichroism was measured and model curves for different secondary structures were fitted by least squares indicating a large component of β-structure. The structure was seen to be homologous with that of the acid proteinase from R. Chinensis and hence with pepsin and chymosin. A rotation function against diffraction data from chymosin crystals confirm this and suggested an approach to the solution of this structure.

REFERENCES

1. Hagemeyer, K., Fawwal, I., and Whitaker, J. R. (1968) J. Dairy Sci. 51, 1916-1930
2. Sardinas, J. L. (1968) Appl. Microbiol. 16, 248-253
3. Williams, D. C., Whitaker, J. R., and Caldwell, P. V. (1972) Arch. Biochem. Biophys. 149, 52-61
4. Moews, P. C., and Bunn, C. W. (1970) J. Mol. Biol. 54, 395-397
5. Jenkins, J. A., Blundell, T. L., Tickle, I. J., and Ungaretti, L. (1975) J. Mol. Biol. 99, 583-590

6. Bunn, C. W., Moews, P. C., and Baumber, M. E. (1971) Proc. R. Soc. London, B178, 245-258

7. Moews, P. C., and Bunn, C. W. (1972) J. Mol. Biol. 68, 389-390

8. Green, M. L. (1975) Biochem. J. 151, 763-764

9. Tickle, I. J. (1975) Acta Crystallogr. B31, 329-330

10. North, A. C. T., Phillips, D. C., and Mathews, F. S. (1968) Acta Crystallogr. 24, 351-359

11. Harding, M. D. (1962) Phil. Thesis, Oxford

12. Matthews, B. (1965) Acta Crystallogr. 20, 230-239

13. Singh, A. K., and Ramaseshan, S. (1966) Acta Crystallogr. 21, 279-280

14. Dodson, E. E., French, S., and Evans, P. (1975) in Anomalous Scattering (Ramaseshan, S. and Abraham, S. C., eds) pp. 423-426 Munkgaard International Publishers Ltd.

15. Blundell, T. L., and Johnson, L. N. (1976) Protein Crystallography, Academic Press, London

16. Dickerson, R. E., Kendrew, J. C., and Strandberg, B. E. (1961) Acta Crystallogr. 14, 1188-1195

17. Blow, D. M., and Matthews, B. W. (1973) Acta Crystallogr. 29, 56-62

18. Blow, D. M., and Crick, F. C. (1959) Acta Crystallogr. 12, 794-802

19. North, A. C. T. (1965) Acta Crystallogr. 18, 212-216

20. Subramanian, E., Swan, I. D. A., Liu, M., Davies, D. R., Jenkins, J. A., Tickle, I. J., and Blundell, T. L. Proc. Natl. Acad. Sci. U.S.A., in press

21. Perlmann, G. E., and Kerwar, G. A. (1973) Arch. Biochem. Biophys. 157, 145-147

22. Grizzuti, K., and Perlmann, G. E. (1969) J. Biol. Chem. 244, 1764-1776

23. Chen, Y. H., Yang, J. T., and Chau, K. H. (1974) Biochemistry 13, 3350-3359

24. Saxena, V. P., and Wetlaufer, D. B. (1971) Proc. Natl. Acad. Sci. U.S.A. 68, 969-972

25. Greenfield, N., and Fasman, G. D. (1969) Biochemistry 8, 4108-4115

26. Wang, T. T., Dorrington, K. J., and Hofmann, T. (1974) Biochem. Biophys. Res. Commun. 57, 865-869

27. Sepulveda, P., Marciniszyn, J., Liu, D., and Tang, J. (1975) J. Biol. Chem. 250, 5082-5088

28. Foltmann, B., and Pedersen, V. B. Private Communication

29. Crowther, R. A. (1972) in Molecular Replacement Method (Rossmann, M. G. ed) Gordon and Breach, New York

PENICILLOPEPSIN: 2.8 A STRUCTURE, ACTIVE SITE CONFORMATION AND MECHANISTIC IMPLICATIONS

I-Nan Hsu, Louis T. J. Delbaere, Michael N. G. James

MRC Group on Protein Structure and Function
University of Alberta, Edmonton, Alberta, T6G 2H7

Theo Hofmann

Department of Biochemistry
University of Toronto
Toronto, Canada M5S 1A8

Penicillopepsin is an acid protease produced by the mold
Penicillium janthinellum at pH's less than 4.1 (1). Enzyme production occurs after the mycelial growth has ceased and sporulation
has begun (2). The specificity and catalytic mechanism of penicillopepsin are very similar to those of porcine pepsin (3). The two
active site aspartic acid residues, Asp-32 and Asp-215, occur in
peptide sequences of at least eight amino acid residues which are
almost identical in penicillopepsin, pepsin and chymosin (1,4-10).
In addition, there is an overall 32% identity of amino acid sequence
between penicillopepsin and porcine pepsin; a tentative sequence
alignment of these two enzymes is given in Table I using this
sequence numbering of porcine pepsin (5). These facts indicate
that the fungal enzyme penicillopepsin is an evolutionary homologue
of the mammalian acid proteases.

CRYSTALLOGRAPHIC DATA

Penicillopepsin crystallizes in the monoclinic system (space
group C2, a = 97.38(8), b = 46.62(3), c = 65.39(6) A) from a 9 mg/ml
solution of the enzyme and 2.5 M Li_2SO_4, 0.1 M sodium acetate at
pH 4.4 (11). A 6 A resolution structure has previously been
described (12). The same data collection techniques were employed
here as were used for the 4.5 A and 2.8 A structures of SGPB (13,14).
Each 2.8 A data set (6700 reflexions) was collected on a single
crystal of the native enzyme and each of the eight heavy-atom derivatives. Heavy atom positions were initially determined from cross-

phased Fourier maps using the 6 A phases, and from difference
Patterson maps. The heavy-atom parameters were adjusted by the
method of least-squares with alternating cycles of protein phase
angle determination. The refinement statistics are summarized in
Table II. The "phasing power" of each of the derivatives is illustrated in Figure 1; all derivatives are phasing to 2.8 A resolution. The overall mean figure of merit, <m>, is 0.90. The native structure
amplitudes and "best" phases (15) of 6127 (91.4% of total) reflexions
were used to compute the electron density map. Several sections
of this map through the active site of the molecule are shown in
stereo in Figure 2.

SECONDARY AND TERTIARY STRUCTURE

The path of the polypeptide chain of penicillopepsin is shown
in the stereo drawing in Figure 3. The molecule has an asymmetric
shape 65 x 49 x 39 A which is similar to those reported for two other
acid proteases from Rhizopus chinensis (16) and Endothia parasitica
(17). A deep cleft separates the two lobes of the molecule (Fig.3)
and the active site Asp-32 and Asp-215 are located in this groove on
the upper and lower lobe, respectively. The upper lobe in Figure 3
is larger and consists of residues -4 to 171, whereas the lower lobe
is composed of the remaining amino acid residues. The molecule of
penicillopepsin has an extended hydrophobic core which has the shape
of a croissant. This croissant is mainly composed of an 18-stranded
mixed β-sheet structure, the largest so far observed in globular
protein structures (18). This sheet starts at the strand of residues
62 to 66 in the upper lobe of the molecule in Figure 3 and twists in
an anticlockwise fashion by 540° in reaching the final strand of
residues 286 to 295 in the lower lobe of the molecule. This sheet
structure is described diagrammatically in Figure 4 with the convention proposed by Richardson (18). In addition to this large β-sheet,
the upper lobe consists of a three-stranded antiparallel β-sheet
formed by residues 71 to 76, 78 to 82 and 105 to 109; the lower lobe
has the single disulfide bridge of the molecule between residues
250 and 283 with residues 256 to 270 forming an antiparallel β-loop.
A helical conformation is adopted by residues 58 to 62, 137 to 141,
225 to 235, and 303 to 309.

ACTIVE SITE

A carboxylate anion and a protonated carboxyl group have been
implicated in the catalytic mechanism of acid proteases (19).
Various diazo compounds react with a single carboxyl group and
completely inactivate pepsin (6,20). Similar inhibition of penicillopepsin with diazoacetylnorleucine ethyl ester (DAN) occurs (1).
The carboxyl group reacting with DAN has been identified as Asp-215
(1,21). More recent studies of pepsin (22,23) with 1,2-epoxy-3-

Table I

Comparison of Tentative Penicillopepsin[a]
Sequence with Porcine Pepsin[b]

Amino acid residues enclosed by solid lines are those regions
of highest homology when compared with McLachlan's (40) sequence
comparison program. Tentative sequence identity with porcine pepsin:
102 residues = 31.7%. The single letter code for the amino acids is
given in Ref.42.

The body of Table I is shown on the two following pages.

Footnotes for Table I

a) The order of the fragments from chemical sequencing (4)
with the residue numbers in this table is as follows: NH_2-
terminal (-4 to 111); PP3 (111 to 121); PP7 (122 to 125);
PP1 (126 to 138); PP2 (139 to 150); central fragment
(151 to 197); PP5 (198 to 207); DAN (211 to 220); PP4
(221 to 231); PP10 (232 to 246); PP11 (247 to 259)[c]; PP6
(260 to 263); PP12 (265 to 299); x-ray map (300 to 302)[c];
C-terminal fragment (303 to 327). Fragments PP8 and PP9
(9 residues) have not been used in this sequence. There is
a major discrepancy between the chemical sequence of PP11
and the electron density map.

b) The sequence numbering is that of porcine pepsin (5).

c) The identification of residues enclosed in parentheses
comes mainly from the electron density and not from chemical
sequencing and therefore these regions are tentative.

Table 1 (continued)

	-4	-3	-2	-1	1	2	3	4	5	6	7	8	9	10	11	12	13	14	15	15	16	17	18	19	20	21	22	23	24
																			A	B									
Pen.	A	A	S	G	V	A	T	N	T	P	T	A	N	-	D	E	E	Y	I	T	P	V	T	I	G	-	-	-	G
Pep.	A	A	A	L	I	G	D	E	P	L	E	N	Y	L	D	T	E	Y	F	-	-	G	T	I	G	I	G	T	P	A

25	26	27	28	29	30	31	32	33	34	35	36	37	38	39	40	41	42	43	44	45	46	47	48	49	50	51	52	53	54
T	T	L	N	F	D	T	S	A	D	L	W	V	F	S	T	E	L	P	A	S	Q	Q	S	G	H	S			
Q	D	F	T	V	I	F	D	T	G	S	S	N	L	W	V	P	S	V	Y	C	S	S	L	A	C	S	D	H	N

55	56	57	58	59	60	61	62	63	64	65	66	67	68	69	70	71	72	73	74	75	76	77	78	79	80	81	82	83
																								A				
V	Y	N	P	S	A	T	G	K	-	E	L	S	G	Y	T	W	S	I	S	Y	G	D	G	S	S	A	S	G	N
Q	F	N	P	D	D	S	S	T	F	E	A	T	S	Q	E	L	S	I	T	Y	G	T	G	S	M	-	T	G	I

84	85	86	87	88	89	90	91	92	93	94	95	96	97	98	99	00	01	02	03	04	05	06	07	08	09	10	11	12	13
															1														
V	F	T	D	S	V	T	V	G	G	V	T	A	H	G	Q	A	V	Q	A	A	Q	Q	I	S	A	Q	F	Q	Q
L	G	Y	D	T	V	Q	V	G	G	I	S	D	T	N	Q	I	F	G	L	S	E	T	E	P	G	S	F	L	Y

14	15	16	17	18	19	20	21	22	23	24	25	26	27	28	29	30	31	32	33	34	35	36	37	38	39	40	41	42
																A												
D	T	N	N	D	G	L	L	G	L	A	F	S	S	I	N	T	V	Q	P	Q	S	Q	T	T	F	F	D	T	V
Y	A	P	F	D	G	I	L	G	L	A	Y	P	S	I	S	A	S	G	A	T	-	P	V	F	D	N	L	W	D

43	44	45	46	47	48	49	50	51	52	53	54	55	56	57	58	59	60	61	62	63	64	65	66	67	68	69	70	71	72
Y	S	S	L	A	Q	P	L	F	A	V	A	L	K	H	Q	-	-	Q	P	G	V	Y	D	F	G	F	I	D	S
Q	G	L	V	S	Q	D	L	F	S	V	Y	L	S	N	D	D	S	G	S	V	V	L	G	G	I	D	S		

	73	74	75	76	77	78	79	80	81	82	83	84	85	86	87	88	89	90	91	92	93	94	95	96	97	98	99	00	01
												A															2		
Pen.	S	K	Y	T	G	S	L	T	Y	T	G	V	D	N	S	Q	G	F	W	S	F	N	V	D	S	Y	T	A	G	S
Pep.	S	Y	Y	T	G	S	L	N	W	V	P	V	–	S	V	E	G	Y	W	Q	I	T	L	D	S	I	T	M	D	G

	02	03	04	05	06	07	08	09	10	11	12	13	14	15	16	17	18	19	20	21	22	23	24	25	26	27	28	29	30	31
Pen.	Q	S	G	D	G	F	–	–	S	G	I	A	D	T	G	T	T	L	L	L	B	D	S	V	V	S	Q	Y		
Pep.	E	T	I	A	C	S	G	G	C	Q	A	I	V	D	T	G	T	S	L	L	T	G	P	T	S	A	I	A	I	N

	32	33	34	35	36	37	38	39	40	41	42	43	44	45	46	47	48	49	50	51	52	53	54	55	56	57	58	59	60
						A																							
Pen.	Y	S	Q	V	S	G	A	Q	Q	D	S	N	A	G	G	Y	V	F	(T	C	S	B	V)	T	B	L	P	V	S	I
Pep.	I	Q	S	D	I	G	A	–	S	E	N	S	D	G	E	M	V	I	S	C	S	S	I	D	S	L	P	D	I	V

	61	62	63	64	65	66	67	68	69	70	71	72	73	74	75	76	77	78	79	80	81	82	83	84	85	86	87	88	89
																					A								
Pen.	S	G	Y	–	T	A	T	V	P	G	S	L	I	N	Y	G	P	S	G	N	G	S	T	C	L	G	G	I	Q	S
Pep.	F	T	I	D	G	V	Q	Y	P	L	S	P	S	A	Y	I	L	Q	D	D	D	S	–	C	T	S	G	F	E	G

	90	91	92	93	94	95	96	97	98	99	00	01	02	03	04	05	06	07	08	09	10	11	12	13	14	15	16	17	18	19
										3																				
Pen.	N	–	–	S	G	I	G	F	(L	I	F)	G	D	I	F	L	K	S	Q	Y	V	V	F	D	S	D	G	P		
Pep.	M	D	V	P	T	S	S	G	E	L	W	I	L	G	D	V	F	I	R	Q	Y	Y	T	V	F	D	R	A	N	N

	20	21	22	23	24	25	26	27
Pen.	Q	L	G	F	A	P	Q	A
Pep.	K	V	G	L	A	P	V	A

Table II

Heavy Atom Refinement Summary

Derivative	Soaking Time (days)	No. Sites[a]	Range A[b] (electrons)	Range B[c] (A^2)	R_c [d]
1 mM $UO_2(C_2H_3O_2)_2$	7	3	11.5-22.6	6.9-18.4	0.512
5 mM $Pt(NH_3)_2Cl_2$	7	2	11.4-43.8	11.4-22.6	0.501
1 mM Pt+1 mM UA	14	4	8.7-21.2	7.4-17.1	0.581
mersalyl (sat'd)	3	7	5.7-28.4	11.6-25.6	0.587
0.5 mM K_2HgI_4	3	5	5.4-27.6	5.3-20.7	0.626
5 mM $K_3UO_2F_5$	21	4	6.7-31.8	10.9-17.8	0.503
phenylmercuric acetate (sat'd)	7	7	6.6-42.7	9.1-17.9	0.569
2.5 mM $IrBr_3$	4	6	4.6-37.4	3.7-43.5	0.653

[a] both the K_2HgI_4 and $IrBr_3$ derivatives had two sites which consisted of trigonal moieties (of HgI_3^- and $IrBr_3$, respectively)

[b] A is the site occupancy on an approximately absolute scale

[c] B is the isotropic temperature factor coefficient

[d] $R_c = \Sigma \left| |F_{PH} - F_P| - |f_H| \right| / \Sigma |F_{PH} - F_P|$ summed over the centric data only

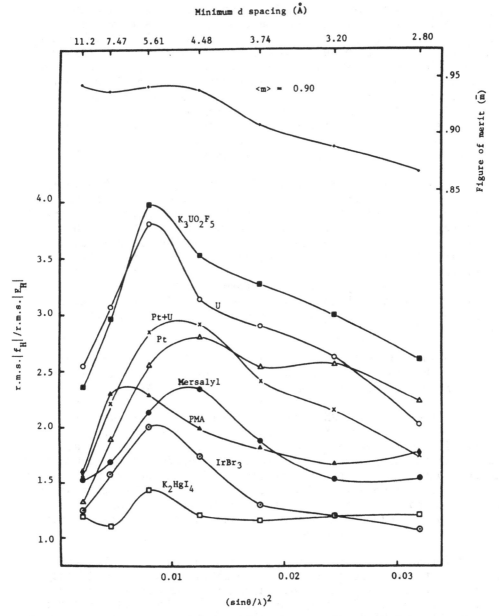

Figure 1. Phasing power of the eight heavy-atom derivatives used for the MIR phase determination of penicillopepsin. The ratio of r.m.s.$|f_H|$ to r.m.s.$|E_H|$ and the figure of merit ($\bar{m}$) are plotted as functions of $\sin^2\theta/\lambda^2$. The derivatives are represented by the following symbols: o, uranyl acetate; Δ, $Pt(NH_3)_2Cl_2$; X, double derivative (Pt+U); •, mersalyl; ▢, K_2HgI_4; ■, $K_3UO_2F_5$; ▲, phenylmercuric acetate; θ, $IrBr_3$.

(p-nitrophenoxy)propane (EPNP) have shown that a second carboxyl
group is involved in the mechanism and this one is from Asp-32.
The two carboxyl groups Asp-32 and Asp-215 of penicillopepsin are
in intimate contact in our structure at 2.8 A resolution (Fig.2).
Direct proof that these two carboxyl groups are the ones involved
in catalysis comes from the following reaction carried out in the
crystals of penicillopepsin. A crystal of penicillopepsin was
soaked for 15 days in a saturated solution of EPNP in normal buffer
at pH 4.4. The 2.8 A diffraction data were collected from this
crystal and a difference electron density map computed from the
amplitudes ($|F_{P\ (EPNP)}| - |F_P|$) and MIR phases (α_p).

The map is extremely clean and shows significant changes in
only two regions. One region involves a non-covalently bound EPNP
molecule in a solvent cavity between two protein molecules. The
other region involves the active site and this portion of the map
is shown in Figure 5. Two molecules of EPNP have reacted covalently
with penicillopepsin, one with the carboxyl group of Asp-32, the
other with the carboxyl group of Asp-215. Also in the region of
the active site is a pronounced conformational change in the position
of Tyr-75, the side chain of which moves towards the EPNP molecule
bound to Asp-32 and is coplanar with the p-nitrophenoxy moiety of
that EPNP.

The work of Hartsuck and Tang (23) on the modification of a
single carboxylate group in the active center of pepsin indicated
that EPNP reacts with the ionized carboxylate of Asp-32. However,
Chen and Tang (22) also isolated a major peptide which had the
sequence Ile-Val-Asp (EPNP) which corresponds to residues 213-215
of the pepsin sequence (5). It is therefore likely in light of the
crystallographic evidence from penicillopepsin that pepsin also reacts with two EPNP molecules. There would appear to be enough room
to accomodate each of the two molecules simultaneously in the active
site of penicillopepsin. However, the possibility exists that the
binding of one EPNP to either Asp-32 or Asp-215 precludes the binding of the second EPNP molecule and that we are observing a statistical disorder in the crystals.

Reference to Figure 2 indicates some of the other important residues in the neighborhood of the two active aspartic acid side
chains. It is clear that Asp-32 and Asp-215 are in very close
contact (preliminary measurements from the Kendrew model indicate
that the two carboxyl groups are approximately coplanar and about
2.6 A separates two of the oxygen atoms). The pH profiles of k_{cat}
or k_{cat}/K_m for pepsin-catalyzed hydrolysis of small synthetic
neutral peptides indicate apparent pK_a-values of 1.0 and 4.7 for
these two reactive carboxyl groups (24). Kitson and Knowles (25)
have pointed out that perturbation of the pKa's of strongly interacting carboxyl groups such as observed in maleic acid could be an
explanation for these two abberrant pK_a values. Indeed, crystal
structure studies (26-28) of the maleate mono-anion and

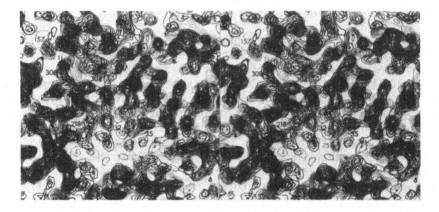

Figure 2. Stereo view of 10 sections (1 A spacing) of the electron density map of penicillopepsin in the catalytic site region. The close contact of Asp-32 and Asp-215 is in the central portion of the figure. The interaction of Asp-304 with main chain Thr-216, Gly-217 is discernible to the left and above Asp-215. The carboxyl group of Asp-32 lies close to this interaction on the opposite side of the main-chain. The relative disposition of the ion-pair Lys-308, Asp-11 to Asp-304 and Tyr-14 can also be seen. The solvent channel of the extended substrate binding groove lies below (in this figure) the active center aspartic acid residues.

maleic acid clearly show that the proton is shared between the two
carboxyl groups (-COO...H...OOC-). Thus, the requirement for a
protonated and an ionized carboxylate group in the active site of
the acid proteases can be satisfied by the disposition of the two
carboxyl groups in the penicillopepsin structure.

In addition to the close contact of Asp-32 and Asp-215, there
are other close contacts to the carboxyl group of Asp-32. These are
shown schematically in Figure 6, which portrays the approximate
spatial distribution of residues in the active site region. The
β-OH group of Ser-35 is within hydrogen bond distance of the carboxylate of Asp-32; also the main chain N-H of Gly-217 is within hydrogen
bonding distance of this group. Thus the spatial environment of the
two reactive aspartic acid side chains is quite different and is
consistent with the low pK_a associated with Asp-32.

Involvement of special peptide bond in the pepsin mechanism
was suggested by Wang (29). The structure of penicillopepsin shows
that there is an involvement of a peptide bond, which is between
Thr-216 and Gly-217 (Figs.2 and 6). However, the involvement is not
in the manner proposed by Wang. Aspartic acid 304 is an internally
buried Asp and its carboxyl group (protonated in this environment)
makes a hydrogen-bonded contact with the carboxyl oxygen atom of
the main chain of Thr-216. This interaction alone does not add
anything to the mechanism of acid protease action, but there are
two groups in this vicinity that are invariant in all acid protease
sequences thus far determined,suggesting a more important role for
Asp-304 and the peptide bond between Thr-216 and Gly-217. Asp-11
and Lys-308 form an ion-pair which is partially buried in the hydrophobic surface of the extended β-structure of penicillopepsin.
Asp-11(Fig.3) is on the extremity of a hairpin β-loop from the
NH_2-terminus and this loop is close to one end of the extended substrate binding groove.

Figure 7 shows a proposed arrangement of the active site residues that could be resulted upon binding of substrate into the extended binding groove of penicillopepsin. The remainder of this
discussion is speculative, but is consistent with many of the facts
known from chemical and enzymatic studies of acid protease catalysis.
We have drawn heavily on the mechanistic proposals of Knowles (30),
Fruton (31), and Hofmann (32),and have incorporated our structural
information into these mechanisms. We propose that substrate binding in the secondary A site, as defined by Takahashi and Hofmann
(33), will disrupt the Lys-308 to Asp-11 ion-pair. A conformational
change is known to occur on the binding of certain tripeptide activa-
tors, such as Leu-Gly-Leu (34) to the secondary A site and this
change could involve a movement of the NH_2-terminal hairpin loop
containing Asp-11. The energy required in breaking this ion-pair is
regained in penicillopepsin by the formation of an alternate ionic
bond of Asp-11 to His-157.

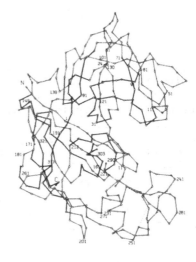

Figure 3. Stereo-drawing of the main polypeptide chain in
penicillopepsin. The circles represent α-carbon positions
and only every 10th residue has been numbered (corresponding
to the numbering of porcine pepsin (5).   The regions of high
amino-acid sequence homology (see Table I) are drawn in dark
lines.   The bilobal feature of the molecule is evident from
this drawing; the active site is located in the groove that
separates the two lobes.

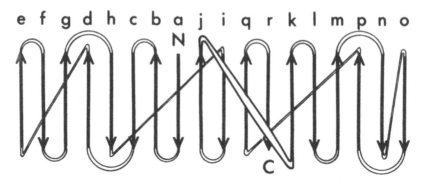

Figure 4. Diagrammatic representation of the 18-stranded mixed
β-pleated sheet structure. This diagram does not attempt to
display the surface topology which can be seen in Figure 3.
Strands of polypeptide chain have been labeled a-r from the N-
to C- terminus. The residue numbers at the start and end of
each strand are: a (-2→8), b (11→18), c (25→33), d (36→41),
e (62→66), f (84→87), g (98→104), h (118→124), i (150→156),
j (161→168), k (179→186), l (189→199), m (202→216), n (218→224),
o (286→295), p (298→303), q (309→316), r (319→325). The sequence
of connection types according to the convention recently published by Richardson (18) is -1, -1, -2, -3X, +1, +2, +5X, -1,
+4X, +1, +1, +2, +1X, -2, -5X, +1.

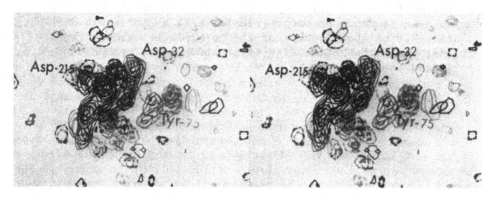

Figure 5. Several sections of the difference electron density
map in the region of the active site after reaction of EPNP
with penicillopepsin, calculated with the amplitudes
$(|F_{P(EPNP)}| - |F_P|)$ and the MIR phases (αp).

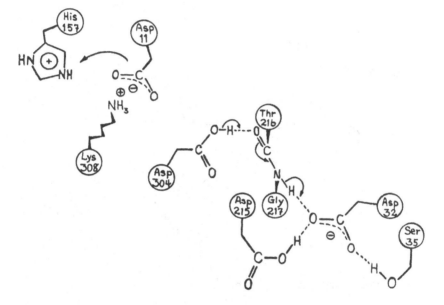

Figure 6. Schematic representation of the catalytically important residues as interpreted from the 2.8 A penicillo-pepsin electron density map (see Fig.2) at pH 4.4.

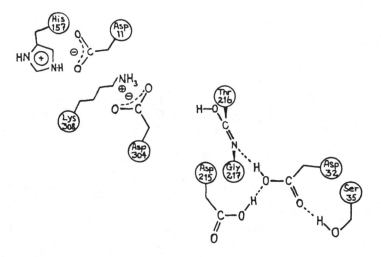

Penicillopepsin   Amino-Intermediate
                            Precursor

Figure 7. Proposal for the alteration of the active site
induced by the binding of substrate in the secondary site A.

Only a minor movement of the side chain of Asp-11 is involved, and His-157 is on the surface and would be positively charged at the pH range in which penicillopepsin is active. There is no homologous positively charged residue corresponding to His-157 in pepsin, but Asp-11 is relatively near the surface and could be solvated by a positively charged ionic species in the case of pepsin.

This conformational change induced by substrate binding to the secondary A site now leaves an almost completely buried positive charge of Lys-308 in the hydrophobic interior. This energetically unfavorable situation is relieved by a subsequent conformational change associated with Asp-304. The carboxyl group hydrogen-bonded to the main chain carbonyl oxygen of Thr-216 would protonate this oxygen and then rotate by $\sim 120^\circ$ about the C_α-C_β bond to form an internally buried ion-pair with Lys-308. The involvement of an arginine side chain in the pepsin mechanism has been suggested by Hartsuck and Tang (23). Their suggestion of Arg-316 is not compatible with the penicillopepsin structure as Arg-316 is a surface residue well-removed from the active site region and, in fact, in penicillopepsin it is a serine. The only basic residue common to both enzymes is that residue at position 308 which in penicillopepsin is a lysine and in pepsin an arginine; we suggest that it is arginine 308 of the pepsin structure which is the residue of importance to the mechanism.

Protonation of the peptide bond joining Thr-216 to Gly-217 would produce a strong acid, as shown in Figure 7. Rather than implicate the imino-hydrogen atom of Gly-217, we have shown protonation of the carboxyl group of Asp-32 and the strongly electrophilic proton would be that proton in Asp-215 (i.e., the carboxyl group of higher $pK_a \sim 4.7$). These conformational changes induced on substrate binding serve to provide the dominant electrophilic component of catalysis as suggested by Knowles (30,35). In addition the rate-limiting step of the overall catalytic process would be associated with these conformational changes that precede the release of products.

MECHANISTIC IMPLICATIONS

In view of the three-dimensional structure of penicillopepsin that we have observed, we are in a position to contribute to an understanding of the mechanism of acid protease catalysis. Any mechanism describing the catalytic event must, of course, be consistent with a number of known facts regarding these enzymes as deduced from chemical and enzymic studies of the cleavage of small synthetic peptide substrates. Knowles has summarized these facts (35) and has presented a possible mechanism for the action of acid proteases (30,35). More recently, Fruton (36) has reviewed the acid protease field and in the following discussion we have attempted to

correlate our structural studies with the more important findings
contained in these several review articles.

Two common features of proteolytic enzymes, whether they belong
to the serine class of proteases or the metalloproteases are: a
nucleophilic group which attacks the carbonyl carbon of the scissile
bond and a proton donor group which transfers a proton to the amide
nitrogen of the peptide bond. We have identified these groups in
the active site of penicillopepsin as Asp-32 and Tyr-75, respectively.
The discussion above, outlined the fact that the side chain of Asp-32
is too buried for it to act as a nucleophile directly; in addition
we feel that large conformational changes required to expose Asp-32
more to the solvent are energetically unfavorable. Too much of the
native structure would have to be disrupted with the concomitant
loss of the proper orientation of the groups in the active site that
we have described above. Even though the β-carboxyl of Asp-215 is
exposed to solvent, it is unlikely that it would act in the role of
the attacking nucleophile due to its higher observed apparent pK_a of
4.7 (25).

The proposed mechanism of acid protease catalysis is illustrated
in Figure 8 in which the bond cleavage step is shown. The scissile
peptide bond of a hypothetical substrate RCO-NHR' is included, but
detailed enzyme-substrate side chain interactions have been omitted
for clarity. We have confirmed via model building experiments on the
Kendrew model of penicillopepsin that substrates such as L-Phe-L-Phe
can be accommodated as shown in Figure 8 and that the substrate's
side chains interact with hydrophobic pockets in the proximity of
the active site.

Before considering the details of the bond cleavage step, we
present a possible sequence of events in the formation of a productive
Michaelis-complex. The extensive antiparallel β-sheet (residues
74-80) which contains Tyr-75 opens outwards (as in the EPNP binding
experiment) to admit the substrate. The substrate binding displaces
two water molecules whereas that near Asp-32 and Ser-35 remains in
its original position. This latter water molecule (see Fig.8A) now
lies between the substrate and the β-carboxyl group of Asp-32. The
antiparallel β-sheet from residues 74-80 moves inward with the
phenolic -OH group of Tyr-75 approaching the amide of the susceptible
peptide bond; the Michaelis complex is now formed as depicted in
Figure 8A.

This proposed mechanism is directly analogous to that advanced
for carboxypeptidase A (37,38). Tyr-75 (in penicillopepsin) corresponds to Tyr-248 (in carboxypeptidase A) whose phenolic -OH group
is in a position to protonate the amide nitrogen of the scissile
bond after a similar conformational change upon substrate binding.
The proton shared between Asp-215 and Asp-32 polarizes the carbonyl
bond in the case of the acid proteases; the electrophilicity of this

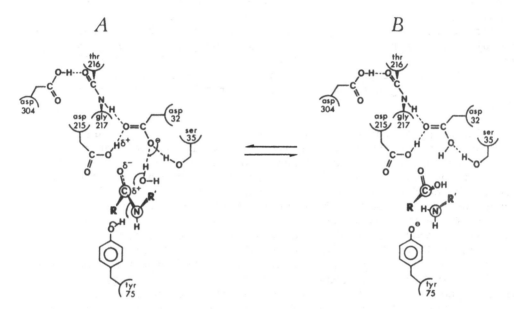

Figure 8. Proposed general base catalytic mechanism of the
acid proteases. "A" shows a schematic drawing of the productive binding mode for substrate RCONHR' in the active site of
penicillopepsin. The conformational change of the "flap"
region brings Tyr-75 to a position from which it can protonate
the amide nitrogen of the scissile bond. Asp-32 attacks the
carbon atom of the polarized $C^{\delta+} \cdots O^{\delta-} \cdots H^{\delta+}$ system through
the bound water molecule W(1). Transpeptidation via acyl or
amino transfer is a function of the enzymes' relative affinity
for the acyl or amino fragment shown in "B".

proton is a function of the nature of the substrate and is enhanced
by the specific enzyme-substrate interactions described previously.
The Zn^{+2} ion in carboxypeptidase A provides the electrophilic component which polarizes the substrate carbonyl bond. Asp-32 of penicillopepsin is analogous to Glu-270 of carboxypeptidase A. The carboxypeptidase mechanism postulates either a direct nucleophilic
attack of the β-carboxyl group of Glu-270 on the carbonyl carbon atom
of the substrate or a general base attack via an activated bound
water molecule. Due to the geometry of the active site of penicillopepsin, and to the inability of workers to trap a covalent enzyme-
substrate intermediate (36), we prefer the general base mechanism of
hydrolysis. Our proposal postulates a ternary complex of enzyme,
acyl-fragment and amino-fragment after cleavage of the scissile bond;
the relative affinity of the acyl or amino moieties for their
respective binding sites determines the order of release of products (39).

Our proposed mechanism is also compatible with the transpeptidation phenomenon which only occurs significantly for acid proteases at
pH values greater than 4.0. Transfer of the proton from the carboxyl
group of one hydrolysis product to the amino group of the other in
Figure 8B could readily occur to produce $R-COO^-$ and H_3^+N-R'. The
transpeptidation products from the incubation of the peptide
Leu-Try-Met with pepsin and penicillopepsin are the amino-trans-
peptidation products Leu-Leu and Leu-Leu-Leu and the acyl-transpep-
tidation products Met-Met and Met-Met-Met (40). In order to produce
the first products Leu-Leu and Leu-Leu-Leu, the tripeptide would bind
initially such that the Leu-Trp bond is cleaved. Exchange of Trp-Met
with a second Leu-Trp-Met is more facile than the exchange of the
bound NH_2-terminal Leu product from the first hydrolysis. Transpeptidation would occur via the microscopic reversibility of the mechanism shown in Figure 8, producing the Leu-Leu-Trp-Met which is subsequently cleaved to yield Trp-Met with the Leu-Leu dipeptide which
can be released as a product or remain bound to the enzyme to undergo
further transpeptidation to yield Leu-Leu-Leu. Fruton (36) has
pointed out that the reaction

$$RCOO^- + H_3^+N-R' \rightleftharpoons RCONHR' + H_2O$$

has a free energy change of around zero at pH values ∿4 i.e. reversal
of the hydrolytic step is equally probable. However, at pH values
∿2 (close to the pH optima of pepsin and penicillopepsin), the carboxyl of the acyl fragment is protonated which precludes transpeptidation of these low pH values.

ACKNOWLEDGMENTS

We are indebted to Dr. G. Lenhert for his system of programs for
the FACS-1 diffractometer (41) and to Dr. M. G. Rossmann for the

phasing program used. This work was funded by the Medical Research Council of Canada in grants to the Protein Structure and Function Group at the University of Alberta (MNGJ) and to T. H. (grant MA-2438).

SUMMARY

The crystal structure of penicillopepsin, an extracellular acid protease isolated from the mold Penicillium janthinellum, has been determined at 2.8 A resolution by the method of multiple isomorphous replacement. The resulting electron density map computed from the native structure factor amplitudes and MIR phases has an overall mean figure of merit of 0.90. The molecule is decidedly nonspherical, with the majority of residues in β-structure. There is an 18-stranded mixed β-sheet which forms the structural core in the region of the active site. This site, identified by the covalent binding of two EPNP molecules to Asp-32 and Asp-215, is located in a deep groove which divides the molecule into two approximately equal lobes. Both aspartic acid residues in the active site are in intimate contact with one another and the carboxyl group of Asp-32 makes two other important hydrogen-bonded contacts: one with Ser-35 and the other with the main chain peptide bond between Thr-216 and Gly-217.

A proposed mechanism for acid protease catalysis is similar in many aspects to that proposed for carboxypeptidase A. The electrophilic component which polarizes the substrate carbonyl bond in the acid proteases is the proton shared between the β-carboxyl groups of Asp-32 and Asp-215. The β-carboxyl group of Asp-32 removes a proton from a water molecule bound between this side chain and the substrate; the resultant OH⁻ attacks the carbonyl carbon atom of the substrate molecule. The phenolic -OH group of Tyr-75 donates its proton to the amide nitrogen of the scissile bond of the substrate.

REFERENCES

1. Sodek, J., and Hofmann, T. (1970) Can. J. Biochem. 48, 1014-1017
2. Mackinlay, A. G., and Wake, R. G. (1971) in Milk Proteins: Chemistry and Molecular Biology Vol. 2 (McKenzie, H. A. ed) pp. 175-215, Academic Press, New York
3. Mains, G., Takahashi, M., Sodek, J., and Hofmann, T. (1971) Can. J. Biochem. 49, 1134-1149
4. Cunningham, A., Wang, H.-M., Jones, S. R., Kurosky, A., Rao, L., Harris, C. I., Rhee, S. H., and Hofmann, T. (1976) Can. J. Biochem. 54, 902-914
5. Tang, J., Sepulveda, P., Marciniszyn, Jr., J., Chen, K. C. S., Huang, W.-Y., Tao, N., Lin, D., and Lanier, J. P. (1973) Proc. Natl. Acad. Sci. U.S.A. 70, 3437-3439

6. Rajogopalan, T. G., Stein, W. H., and Moore, S. J. (1966) J.
    Biol. Chem. 241, 4295-4297
7. Bayliss, R. S., Knowles, J. R., and Wybrandt, G. B. (1969)
    Biochem. J. 113, 377-386
8. Meitner, P. A. (1971) Biochem. J. 124, 673-676
9. Foltmann, B., Kauffman, D., Parl, M., and Andersen, P. M. (1973)
    Netherlands Milk Dairy J. 27, 165-175
10. Takahashi, K., Mizobe, F., and Chang, W. J. (1972) J. Biochem.
    (Tokyo) 71, 161-164
11. Camerman, N., Hofmann, T., Jones, S., and Nyburg, S. C. (1969)
    J. Mol. Biol. 44, 569-570
12. Hsu, I-N., Hofmann, T., Nyburg, S. C., and James, M. N. G. (1976)
    Biochem. Biophys. Res. Commun. 72, 363-368
13. Codding, P. W., Delbaere, L. T. J., Hayakawa, K., Hutcheon,
    W. L. B., James, M. N. G., and Jurasek, L. (1974) Can. J.
    Biochem. 52, 208-220
14. Delbaere, L. T. J., Hutcheon, W. L. B., James, M. N. G., and
    Thiessen, W. E. (1975) Nature 257, 758-763
15. Blow, D. M., and Crick, F. H. C. (1959) Acta Crystallogr. 12,
    794-802
16. Subramanian, E., Swan, I. D. A., and Davies, D. R. (1976)
    Biochem. Biophys. Res. Commun. 68, 875-880
17. Jenkins, J. A., Blundell, T. L., Tickle, I. J., and Ungaretti, L.
    (1975) J. Mol. Biol. 99, 583-590
18. Richardson, J. S. (1976) Proc. Natl. Acad. Sci. U.S.A. 73,
    2619-2623
19. Fruton, J. S. (1970) Adv. Enzymol. 33, 401-443
20. Delpierre, G. R., and Fruton, J. S. (1966) Proc. Natl. Acad. Sci.
    U.S.A. 56, 1817-1822
21. Knowles, J. R., and Wybrandt, G. B. (1968) FEBS Lett. 1, 211
22. Chen, K. C. S., and Tang, J. (1972) J. Biol. Chem. 247, 2566-2574
23. Hartsuck, J. A., and Tang, J. (1972) J. Biol.Chem. 247, 2575-2580
24. Cornish-Bowden, A. J., and Knowles, J. R. (1969) Biochem. J.
    113, 353-362
25. Kitson, T. M., and Knowles, J. R. (1971) FEBS Lett. 16, 337-338
26. James, M. N. G., and Williams, G. J. B. (1971) J. Med. Chem.
    14, 670-675
27. James, M. N. G., and Williams, G. J. B. (1974) Can. J. Chem.
    52, 1872-1879
28. James, M. N. G., and Williams, G. J. B. (1974) Acta Crystallogr.
    B30, 1249-1257
29. Wang, J. H. (1970) in Structure-Function Relationships of Proteo-
    lytic Enzymes (Desnuelle, P., Neurath, H. and Otteson, M.,eds)
    pp. 251-252, Munksgaard, Copenhagen
30. Knowles, J. R. (1970) Philos. Trans. R. Soc. London Ser B 257,
    135-146
31. Delpierre, C. R., and Fruton, J. S. (1965) Proc. Natl. Acad.
    Sci. U.S.A. 54, 1161-1167
32. Hofmann, T. (1974) Adv. Chem. Series 136, 146-185
33. Takahashi, M., and Hofmann, T. (1975) Biochem. J. 147, 549-563

34. Wang, T.-T., and Hofmann, T. (1976) Biochem. J. 153, 701-712
35. Knowles, J. R., Bayliss, R. S., Cornish-Bowden, A. J., Greenwell,
    P., Kitson, T. M., Sharp, H. C., and Wybrandt, G. B. (1970) in
    Structure-Function Relationships of Proteolytic Enzymes"
    (Desnuelle, P., Neurath, H., and Otteson, M. eds) pp. 237-250,
    Munksgaard, Copenhagen
36. Fruton, J. S. (1976) in Advances in Enzymology (Meister, A., ed)
    pp. 1-36, Interscience, John Wiley and Sons, New York
37. Lipscomb, W. N., Hartsuck, J. A., Reeke, G. N., Quiocho, F. A.,
    Bethge, P. H., Ludwig, M. L., Steitz, T. A., Muirhead, H., and
    Coppola, J. C. (1968) Brookhaven Symp. Biol. 21, 24-90
38. Quiocho, F. A., and Lipscomb, W. N. (1971) in Advances Protein
    Chemistry (Afinsen, C. B., Edsall, J. T., and Richards, F. M.
    eds) pp. 1-78, Academic Press, New York
39. Newmark, A. K., and Knowles, J. R. (1975) J. Amer. Chem. Soc.
    97, 3557-3559
40. Wang, T.-T., and Hofmann, T. (1976) Biochem. J. 153, 691-699
41. Lenhert, P. G. L. (1975) J. Appl. Crystallogr. 8, 568-570
42. Biochem. J. (1969) 113, 1-4

MECHANISM OF PEPSINOGEN ACTIVATION

INTRAMOLECULAR ACTIVATION OF PEPSINOGEN

Jean A. Hartsuck, Joseph Marciniszyn, Jr.,
Jung San Huang, and Jordan Tang

Laboratory of Protein Studies
Oklahoma Medical Research Foundation
and
Department of Biochemistry and Molecular Biology
University of Oklahoma Health Sciences Center
Oklahoma City, Oklahoma 73104

Porcine pepsinogen comprises a single polypeptide chain of 371 residues. The NH_2-terminal sequence is Leu-Val- (1,2). After activation, pepsin also is a single polypeptide chain. The dominant pepsin species results from removal of 44 residues from the NH_2-terminus of the pepsin so that the NH_2-terminal sequence is Ile-Gly- (3). Other NH_2-terminal sequences have been observed (3,4,5). Such heterogeneity is mainly the result of removal of a different number of residues from the NH_2-terminus of pepsinogen. Under certain activation conditions, however, cleavage of internal peptide bonds also occurs, so that the resulting mixture contains pepsin molecules, which are not a single polypeptide chain (5).

The first kinetic study of pepsinogen activation was made by Herriott (6), who showed that pepsinogen is capable of spontaneous activation if it is subjected to an acidic pH of 6 or less. At pH 4.6, 25°, an 8.75 mg/ml pepsinogen solution is half activated in about 6 hours. The observed data for pepsin concentration as a function of time fit an autocatalytic rate equation of the form:

$$- \frac{d[Pgn]}{dt} = k_2 [Pgn] \left([Pep_f] - [Pgn] \right) \quad (1)$$

where $[Pgn]$ is the pepsinogen concentration at time t, $[Pep_f]$ is the final pepsin concentration, and k_2 is the second-order rate constant. Therefore, these studies implied that the mechanism of pepsinogen activation is autocatalytic; i.e. it is a bimolecular process in which pepsin cleaves pepsinogen. Herriott also reported a dramatic increase in activation rate if the pH of activation is lowered.

Rajagopalan, Moore, and Stein (5) showed that if pepsin is prepared from pepsinogen by activating an 8.6 mg/ml pepsinogen solution
at pH 2, 2^O, for 20 min, then stopping the reaction by raising the
pH to 4.4, and separating the activation peptides in sulfoethyl
Sephadex C-25 column chromatography, the resulting pepsin is homogeneous and possesses an NH_2-terminal Ile-Gly- sequence. Their study
also showed that not nearly so homogeneous a pepsin is produced if
activation is carried out at higher pH, viz. pH 3 or pH 3.9.

Sepharose-bound pepsinogen was employed by Bustin and Conway-
Jacobs (7) to show that unimolecular pepsinogen activation is possible. In their experiments, activation of immobilized pepsinogen by
exposure to acid was confirmed by the generation of isoleucine NH_2-
terminus in the bound pepsinogen as well as the production of proteolytic capability in the enzyme, which was attached to the column.
Subsequently, McPhie (8) used the appearance of a spectral change to
kinetically study pepsinogen activation above pH 3.75. These data
were interpreted as consistent with mixed activation, i.e., the
existence of both unimolecular and bimolecular activation processes.

KINETIC PROOF OF INTRAMOLECULAR PEPSINOGEN ACTIVATION

On the basis of the experiments described above, we undertook to
measure the pepsinogen activation rate at low pH. The detailed
description of these experiments has been published elsewhere (9) and
only the major results will be recapitulated here. The assay system
we used was patterned after that of Herriott (6) and takes advantage
of the fact that pepsin is rapidly denatured at pH 8.5 where pepsinogen is stable (10). For each time point in the pepsinogen activation
assay, the pH of a pepsinogen solution was lowered rapidly to the
activation pH. After an appropriate time interval, the solution was
diluted as the pH was raised to 8.5. Preliminary experiments determined the amounts of acid and base required to achieve the desired
pH's. After incubation to destroy all pepsin which had been formed,
the remaining pepsinogen was activated at pH 2 and assayed with hemoglobin as substrate. Optical density of trichloroacetic acid-soluble
peptides was determined. This system measures pepsinogen remaining
after activation and treats as pepsin all species which are denatured
by 20 min incubation at pH 8.5.

The results of these experiments confirmed that, with a judicious
choice of protein concentration and pH, first-order kinetics could be
observed. This is verified by the linear semilog plots of pepsinogen
concentration as a function of time, which are shown in Figure 1,
and by the lack of dependence of the first-order rate constant on
protein concentration, which is shown in Table I.

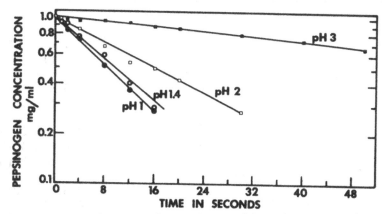

Figure 1. Typical semilog plots of pepsinogen activation at different pH values. Protein concentration in all instances is 1 mg/ml.

Table I

First-Order Pepsinogen Activation Rate Constants

Pepsinogen Concentration	pH	k_1^a
mg/ml		min^{-1}
1.0	1.0	$4.1 \pm 0.7(9)$[b]
0.5	1.0	3.8 (1)
1.0	1.4	3.9 ± 0.3 (10)
0.5	1.4	4.6 ± 0.1 (2)
1.0	2.0	2.6 ± 0.2 (22)
0.5	2.0	2.9 ± 0.4 (5)
0.1	2.0	2.7 ± 0.2 (3)
1.0	3.0	0.39 ± 0.05 (7)
0.5	3.0	0.53 (1)
1.0	4.0	0.008 ± 0.0005 (2)[c]
0.5	4.0	0.008 (1)[c]

[a]First-order rate constant which is evaluated from the least squares fit of the semilog plot except at pH 4.

[b]The form of the data is arithmetic mean ± root mean square deviation (number of observations).

[c]Rate constant was evaluated by curve fitting to the mixed reaction equation (2) (see text).

At pH 4, the pepsinogen concentration versus time did not decline exponentially. Consequently, we assumed that both the first-
order and autocatalytic activation mechanisms were operative. The
rate equation for this reaction is:

$$- \frac{d[Pgn]}{dt} = k_1 [Pgn] + k_2 [Pgn] \; ([Pep_f] - [Pgn]) \quad (2)$$

where $[Pgn]$, $[Pep_f]$, k_2 and t are the same as in equation (1), and
k_1 is the first-order rate constant. The integrated form of equation
(2) fit the pH 4 data well and this curve fitting was used to evaluate both k_1 and k_2.

The addition of preformed pepsin to a pepsinogen solution that
was to be activated at pH's below 4 enhance the activation rate.
Initial rate data in the presence of preformed pepsin were fit to
equation (2) in order to evaluate k_2 at low pH. In this treatment,
k_1 was taken as the value determined in the absence of preformed
pepsin.

The pH profiles of the values of k_1 and k_2 are shown in Figure 2.
As the pH is lowered, k_1 rises whereas k_2 has a maximum at pH 3.
This pH dependence is consistent with a predominance of autocatalytic
kinetics at pH 4 and intramolecular activation below pH 3 as long as
the protein concentration is 1 mg/ml or less.

MECHANISM OF INTRAMOLECULAR ACTIVATION OF PEPSINOGEN

On the basis of spectral changes which occurred when pepsinogen
was subjected to low pH, McPhie (8) suggested that pepsinogen is
capable of a rapid, reversible conformation change. Subsequent work
in our laboratory (4) confirmed the existence of the spectral change,
but showed its interpretation to be more complex than originally
thought. If the pH is lowered under conditions where one expects
partial activation, the spectral change disappears when the solution
is returned to neutrality. If incubation at neutral pH is continued,
however, a similar difference spectrum reappears. We interpreted
these results to mean that after short activation the solution contains pepsinogen as well as pepsin and derived peptides. When the pH
is raised, the pepsin and peptides bind to each other and the spectrum of the complex is similar to that of pepsinogen. Upon incubation at neutral pH, the pepsin-peptide complex dissociates and the
spectral change reappears. Difference spectra which show these
changes are demonstrated in Figure 3. Note that a comparison of
spectra A and D shows D to be of lesser magnitude. This may be explained by presuming that all of the pepsinogen has undergone conformational change in solution A, but solution D contains some native
pepsinogen which has reverted to its original conformation. Because
a change in the fluorescence spectrum precedes the appearance of

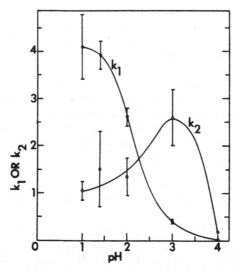

Figure 2. pH dependence of the first- and second-order rate constants, k_1 and k_2. The units for k_1 are $(min)^{-1}$, for k_2 are $(min \times mg\ per\ ml)^{-1}$. The vertical lines indicated root mean square deviations in the rate constant determinations.

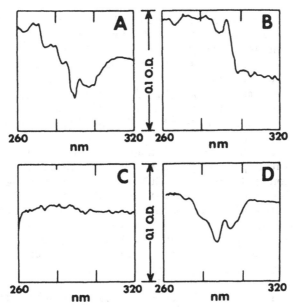

Figure 3. Difference spectra of pepsinogen solutions. A, sample, 20 µM pepsinogen during 1.5-min activation at pH 2.85; reference, 20 µM pepsinogen, pH 7.5. C, same conditions as B but 1 min after neutralization of sample to pH 7.3. D, sample, 20 µM pepsinogen incubated 12 min at pH 8.5 after 1.5-min activation at pH 2.85; reference, 20 µM pepsinogen, pH 8.5.

activity during chicken pepsinogen activation, Bohak (11) concluded
that pepsinogen activation is a two-step process. With this work as
a basis, we undertook to demonstrate the existence of a pepsinogen
intermediate during activation (12).

First, the presence of a pepsinogen intermediate was indicated
because pepstatin inhibited intramolecular pepsinogen activation. In
a typical experiment, conditions which led to 57% activation gave
rise to only 10% activation when the solution was saturated with
pepstatin. In these studies the production of pepsin was evaluated
from the appearance of its Ile-Gly- NH_2-terminal sequence (12).

In addition, a pepstatin-aminohexyl-Sepharose 4B column was employed to demonstrate that at low pH there exists a pepsinogen species
whose conformation apparently is different from that of pepsinogen at
neutral pH (12). Figure 4A shows that pepsinogen at pH 5.6, 1N NaCl,
was not bound by the column. Under similar conditions, pepsin was
retained by the column, Figure 4B. Figure 4C describes an experiment
in which pepsinogen, in neutral pH, was added in batch to the pepstatin-aminohexyl-Sepharose 4B at pH 2.2 at 4^0. This preparation
then was poured into a column. No protein could be eluted with the
buffer used in frame A where pepsinogen did not bind. A low salt,
pH 8 solution eluted pepsiongen with full proteolytic capability.
This protein accounted for about 30% of that added originally. The
remaining pepsinogen was eluted with 1N NaCl, pH 8. This protein had
only 5% of its original specific activity. In all instances the
identity of the eluted protein was established by NH_2-terminal
sequence analysis and polyacrylamide electrophoresis. We interpreted
this experiment to mean that pepsinogen had undergone a conformational change when the pH was lowered. After that change, the active
site area in pepsinogen was capable of binding pepstatin; i.e. it
resembled the active site of pepsin. This pepsinogen binding to pepstatin prohibited the pepsinogen cleavage which accompanies pepsin
formation.

Since pepstatin binds so tightly to pepsin, it was not suitable
for kinetic studies of the inhibition of pepsinogen activation. Consequently, we chose to observe the activation of pepsinogen in the
presence of various concentrations of globin, the protein moiety
from hemoglobin. For these assays, pepsinogen was allowed to activate at pH 2 in the presence of a known concentration of globin.
After an appropriate time interval, the pH was raised to 8.5 so that
alkaline denaturation of all pepsin present would occur. Next, the
remaining pepsinogen was fully activated at pH 2, and then globin
substrate was added so that the pepsin formed in the second activation could be quantified in the usual assay. In this system, a blank
was prepared by adding trichloroacetic acid before the final globin
addition. Use of this blank appropriately subtracted from the total
optical density due to those soluble peptides formed during the two
activation steps. The results of these studies are shown in Figure 5.
These semilog plots are linear for all globin concentrations.

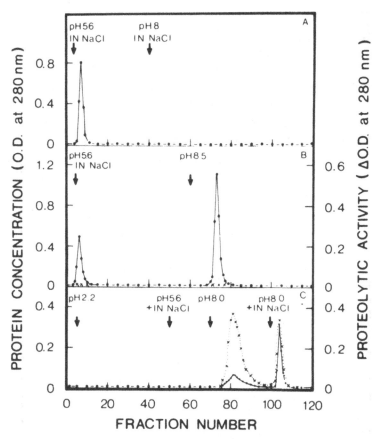

Figure 4. Elution patterns on an affinity column of pepstatin-aminohexyl-Sepharose 4B. Each fraction contains 3 ml of effluent. The solid line and the dotted line are, respectively, the protein concentration and proteolytic activity. A, pepsinogen in 1 M sodium acetate buffer, pH 5.6, with 1 N NaCl was not retained by the affinity column and emerged in the breakthrough peak. B, pepsin in 1 M sodium acetate buffer, pH 5.6 with 1 N NaCl was retained by the column and was eluted in 0.05 M sodium phosphate buffer, pH 8.5, as alkaline inactivated pepsin; the smaller breakthrough peak represents some inactive pepsin in the commercial preparation. C, pepsinogen in 0.1 M glycine/HCl buffer, pH 2.2, was retained by the affinity column and could not be recovered by eluting with 0.1 M sodium acetate buffer, pH 5.6, with 1 N NaCl. Two peaks were eluted subsequently with 0.01 M sodium phosphate buffer, pH 8, and the same buffer containing 1 N NaCl.

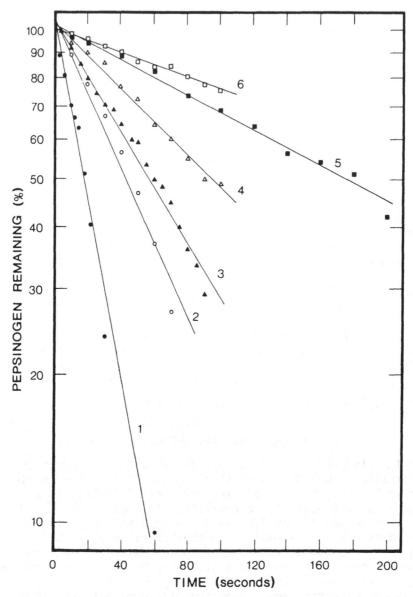

Figure 5. Disappearance of pepsinogen during the intramolecular activation as a function of time in the presence of various concentrations of globin. The numbers next to the lines are the experiment numbers. At zero time, 37°, pepsinogen in pH 5.3 was rapidly acidified to pH 2.0 to a concentration of 1×10^{-6}M. Globin concentrations for experiments 1 through 6 are: 0, 1.56, 3.42, 7.11, 16.16, and 28.41 $\times 10^{-5}$M.

The scheme shown in Figure 6 represents the minimum number of steps
required for pepsinogen activation. To be consistent with the observed first-order kinetics, the equilibria described by the constants $K_{i,G}$ and K_C must be established rapidly with respect to the
bond cleavage k_{1st}. Unfortunately, direct evaluation of the constants, $K_{i,G}$, K_C and k_{1st} is not possible from these kinetic data.
Even so, the linear nature of the plot of the reciprocal of the
apparent pepsinogen decay constant (from Fig.5) versus the globin
concentration (Fig.7) confirms the competitive inhibition of pepsinogen activation by globin. This competitiveness is implied by the
scheme of Figure 6. The detailed interpretation of these data has
been presented previously (12). Suffice it to say that we concluded
that globin is reversibly bound by pepsinogen. The pepsinogen which
is bound to globin cannot undergo activation.

As an adjunct to the experiments just described, we studied the
activation of pepsinogen in the presence of globin in a pH stat. In
this case, the rate of consumption of acid was deemed proportional to
the pepsin present since pepsin when formed began hydrolyzing globin.
In these experiments the same activation rate was observed as in the
experiments where the remaining pepsinogen was activated and assayed.
This result indicates that pepsinogen itself is incapable of globin
hydrolysis and that, to the accuracy of these experiments, the
appearance of denaturable pepsin and proteolytic capability are
simultaneous.

ACTIVATION OF ACETYLPEPSINOGEN

Preliminary experiments on the activation of acetylpepsinogen
indicate that after acetylation of pepsinogen with acetic anhydride
the activation rate is changed (13). In our experiment, presumably,
the major change in the pepsinogen is the acetylation of the lysine
residues, which are located predominantly in the activation peptide.
The acetylpepsinogen was prepared by reacting 4.8 ml of a 6.1 mg/ml
solution of pepsinogen at pH 7 with 70 μl of acetic anhydride added
in increments of 20, 20, 20 and 10 μl. The pH was maintained between
6.5 and 7.5 throughout the acetylation and was allowed to stabilize
between acetic anhydride additions. In the preliminary experiments,
it was found that the potential activity of pepsinogen continued to
decline during the acetylation. In addition, acetylated pepsinogen
loses its potential proteolytic activity rapidly at pH 7-8 (unpublished results). The rate of activation could be measured on the
partially acetylated pepsinogen, however, after a rapid desalting
process. Figure 8 shows a representative rate curve generated by the
activation of acetylpepsinogen at 0.044 mg/ml, pH 2. In this experiment about 20% of the original potential proteolytic activity was
present after acetylation and desalting. Our interpretation of this
data is that the sample contains 22% unmodified pepsinogen which

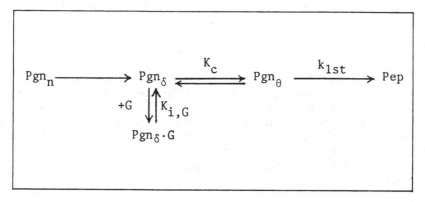

Figure 6. Scheme for the activation of pepsinogen. Pgn_n, Pgn_δ
and Pgn_θ are different conformational states of pepsinogen, Pep
is pepsin; and G is globin. The choice of the Greek letters to
represent the various forms of pepsinogen was dictated by the
following considerations: Pgn_n represents native pepsinogen.
The Pgn_δ species is one in which the active site region is exposed and consequently the activation peptide is not thought to
be adjacent to the bulk of the molecule. The Pgn_θ species has
the activation peptide in the active site and occurs just before
intramolecular cleavage. At some points these species are referred to simply as δ or θ, or as intermediate δ, for example.

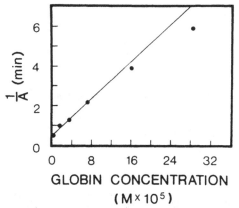

Figure 7. Plot of reciprocal of the pepsinogen decay constant
determined from the data of Figure 5, versus globin concentration. The line drawn is a least squares line calculated from
the four lowest globin concentrations.

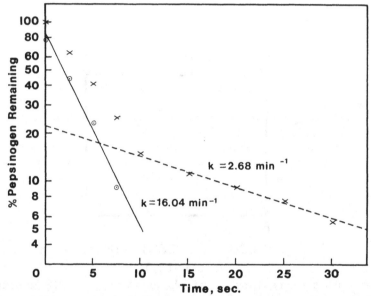

Figure 8. Semilog plot of acetylpepsinogen activation at pH 2. The x's are the observed data points. This curve was resolved into two first-order reactions. The time points greater than 10 s were used to define the slow process. The ⊙'s were calculated for the fast reaction by subtracting the slow contribution from the observed data points. Lines were fit to the appropriate points by least squares analysis.

activates with the first-order constant of 2.7 min^{-1}; this represents
about 5% of the original pepsinogen. The acetylpepsinogen is activated, also unimolecularly, with a rapid rate constant viz. 17.3
min^{-1}. These results seem to imply that the bond cleaved during
activation has become more labile with acetylation of the lysine
side chains of pepsinogen.

DISCUSSION

As a result of our kinetic studies on the activation of porcine
pepsinogen, we have concluded that the slowest step in pepsinogen
activation is an intramolecular bond cleavage. It is important to
note that since the amount of pepsin formed is measured by quantitat-
ing the active pepsinogen after base denaturation, our only characterization of the pepsin species is its loss of activity upon exposure to a basic solution. We have not chemically characterized the
pepsin which is formed and so we cannot identify the bond which is
cleaved by the first-order process. Trapping pepsinogen with pepstatin, which is a transition state analog inhibitor of acid proteases
(see Chapter 12), and the competitive inhibition of intramolecular
pepsinogen activation by globin, a protein substrate of pepsin, indicate that there is a form of pepsinogen which has an exposed active
site. We call this form intermediate δ (see Fig.6). To fit the observed inhibition kinetics, δ must be formed before the rate limiting,
intramolecular bond cleavage. On the basis of our previous arguments
(9), as well as the competitive inhibition of activation by globin
(12), we presume that the same active site binds pepstatin as well as
substrates such as globin in both pepsinogen and pepsin and is responsible for the intramolecular bond cleavage. If this is so, then
the pepsinogen species n, δ and θ must all exist. A diagrammatic
representation of our suggested mechanism of pepsinogen activation is
shown in Figure 9. Both forms φ and α would be base denaturable and
therefore called pepsin. Additional evidence for a pepsinogen conformational change before bond cleavage has been obtained from changes
in the U.V. difference spectrum measured with the stopped-flow
technique (14).

Kay and Dykes (see Chapter 7) have presented evidence which
suggests that the first bond cleavage in the generation of pepsin
produces a protein species whose amino-terminus corresponds to residue 17 of pepsinogen rather than residue 45, which is ultimately the
predominant pepsin NH$_2$-terminus after the unimolecular process.
These pepsinogen activation experiments were done in the presence of
pepstatin. In the absence of pepstatin the appearance of Ile-Gly-
pepsin, with 44 NH$_2$-terminal residues of the zymogen removed, can be
correlated well with the first-order activation rate process as
measured in our kinetic studies (4). If the existence of the protein
species with 16 residues removed is substantiated, especially in the
absence of pepstatin, then the best reconcilliation of all of the

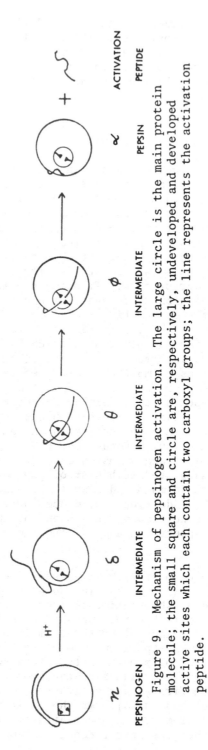

Figure 9. Mechanism of pepsinogen activation. The large circle is the main protein molecule; the small square and circle are, respectively, undeveloped and developed active sites which each contain two carboxyl groups; the line represents the activation peptide.

data seems to be as follows:

The bond cleavage which takes place between intermediates θ and
φ, Figure 9, generates first a protein with 16 residues removed.
Subsequent degradation generates the species α with a predominant
NH_2-terminus at residues 45 of pepsinogen, i.e. Ile-Gly- sequence.
If all species from which at least 16 residues had been removed are
base denaturable, then this interpretation is consistent with our
kinetic experiments. In this instance the removal of the peptide
comprising residues 17-44 may well be by one or more bimolecular
processes. If one assumes a strict structural homology between
porcine pepsin and the fungal proteases from Penicillium janthinellum,
Rhizopus chinensis, and Endothia parasitica, (see Chapters 2,3,4, and
5) then this suggestion of subsequent bimolecular degradation is
attractive since the NH_2-termini of these enzymes are not located
near their active centers.

Prochymosin, from calf stomach, has a primary structure which is
homologous to that of porcine pepsinogen (see Chapter 1). In experiments similar to those of Bustin and Conway-Jacobs (7), Pedersen and
Foltmann (15) now have shown prochymosin to be activated intramolecu-
larly. The only other acid proteases known to have zymogens are gastricsin (16), the protease from human seminal plasma (see Chapter 20)
and kidney renin. The first two of these may well activate by a
mechanism similar to that of pepsinogen since in all known properties
they are similar to pepsinogen. Prorenin, molecular weight 140,000,
certainly possesses a different structure (see Chapter 14 and 15).

ACKNOWLEDGMENT

This work was supported by Research Grants AM-01107 and AM-06487
from the National Institutes of Health.

SUMMARY

Two pathways for pepsinogen activation have been demonstrated.
Intramolecular activation, which is kinetically first-order, pre-
dominates over the autocatalytic pathway if the pH is below 3 and the
protein concentration is less than 1 mg/ml during activation. Intramolecular pepsinogen activation is inhibited either by pepstatin, a
potent pepsin inhibitor, or by purified globin from hemoglobin, a good
pepsin substrate. Also, pepsinogen at pH 2 can be bound to a pepstatin-Sepharose column and recovered as native zymogen upon elution
in pH 8 buffer. Kinetic studies of the globin inhibition of pepsinogen activation show that globin binds to a pepsinogen intermediate.
This interaction gives rise to competitive inhibition of intramolecular pepsinogen activation. The evidence presented in this paper suggests that pepsinogen is converted rapidly upon acidification to the
pepsinogen intermediate δ.

In the absence of an inhibitor, the intermediate undergoes conformational change to bind the activation peptide portion of this same pepsinogen molecule in the active center to form an intramolecular enzyme-substrate complex (intermediate θ). This is followed by the intramolecular hydrolysis of a peptide bond which results in a base-denaturable pepsin species. Intermediate δ apparently does not activate another pepsinogen molecule via an intermolecular process. Neither does intermediate δ hydrolyze globin substrate.

REFERENCES

1. Ong, E. B., and Perlmann, G. E. (1968) J. Biol. Chem. 243, 6104-6109
2. Pedersen, V. B., and Foltmann, B. (1973) FEBS Lett. 35, 255-256
3. Sepulveda, P., Marciniszyn, J., Jr., Liu, D., and Tang, J. (1975) J. Biol. Chem. 250, 5082-5088
4. Sanny, C. G., Hartsuck, J. A., and Tang, J. (1975) J. Biol. Chem. 250, 2635-2639
5. Rajagopalan, T. G., Moore, S., and Stein, W. H. (1966) J. Biol. Chem. 241, 4940-4950
6. Herriott, R. M. (1938) J. Gen. Physiol. 21, 501-540
7. Bustin, M., and Conway-Jacobs, A. (1971) J. Biol. Chem. 246, 615-620
8. McPhie, P. (1972) J. Biol. Chem. 247, 4277-4281
9. Al-Janabi, J., Hartsuck, J. A., and Tang, J. (1972) J. Biol. Chem. 247, 4628-4632
10. Bovey, F. A., and Yanari, S. S. (1960) The Enzymes (Boyer, P. D., Lardy, P. D., and Myrbäck, K. eds) 2nd edition, Vol.4, pp. 63-92
11. Bohak, Z. (1973) Eur. J. Biochem. 32, 547-554
12. Marciniszyn, J., Jr., Huang, J. S., Hartsuck, J. A., and Tang, J. (1976) J. Biol. Chem. 251, 7095-7102
13. Al-Janabi, J., Hartsuck, J. A., and Tang, J., unpublished results
14. Koga, D., and Hayashi, K. (1976) J. Biochem. 79, 549-558
15. Pedersen, V. B. (1976) Abstr. X[th] Int. Union Biochem. Meet. Hamburg, Abstr. No. 04-3-319
16. Tang, J. (1970) Methods Enzymol. 19, 406-421

THE FIRST CLEAVAGE SITE IN PEPSINOGEN ACTIVATION

John Kay and Colin W. Dykes

Department of Biochemistry

University College, Cardiff, Wales

The carboxyl proteases which are responsible for initiating gastric digestion of proteins in most mammals are synthesized in precursor form. The pepsinogens from pig, cow and calf prochymosin, in particular, are well-characterized proteins (see Foltmann and Pedersen, Chapter 1 in this volume). These zymogens convert themselves into the more active forms of their respective enzymes by undergoing limited proteolysis in which an NH_2-terminal 'activation' segment is cleaved off (Fig.1). In porcine pepsinogen, the activation follows an intramolecular mechanism at pH 3 and below, whereas above this pH value the reaction involves the intermolecular action of newly formed pepsin interacting with pepsinogen to form even more pepsin (1-3).

For the intramolecular activation, the pepsin active site (involving at least two well-characterized B-carboxyl groups of aspartic acid residues 32 and 215 {4}) inherent in the zymogen for its own activation must gain exposure by catalyzing the removal of the activation segment. The complete sequences of these activation segments are known (5,6) and Figure 1 shows that slightly more than one tenth of each molecule is trimmed off during conversions. For porcine pepsinogen, 44 amino acids are removed and Ile-45 is generated as the amino-terminal residue of porcine pepsin. None of the activation segments, however, has ever been isolated as one intact polypeptide, and the identity of the first peptide bond to hydrolyze on activation is still uncertain.

Two reaction pathways are possible for the intramolecular transformation:
A. Pepsinogen converts itself into pepsin directly, liberating the activation segment as an intact polypeptide. In practice,

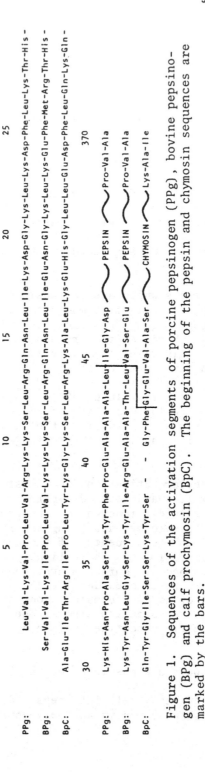

Figure 1. Sequences of the activation segments of porcine pepsinogen (PPg), bovine pepsinogen (BPg) and calf prochymosin (BpC). The beginning of the pepsin and chymosin sequences are marked by the bars.

this has never been detected, and it has been assumed that this
is due to digestion of the polypeptide into smaller pieces by
the active pepsin formed.

B. Pepsinogen is converted into pepsin by a sequential process
in which smaller peptides are removed successively from the
NH_2-terminus until all 44 residues in the activation segment
have been removed, e.g.

Pepsinogen ────────────────→ Peptide a + Intermediate Protein A

Intermediate Protein A ──────→ Peptide b + Intermediate Protein B

Intermediate Protein B ──────→ Peptide n + Pepsin

The former possibility has been widely accepted as correct with
little experimental substantiation, but both possibilities, in addition to generating active pepsin, account for the number of activation peptides usually observed.

It should be possible to distinguish between the two hypotheses
if the active product could be trapped into an inactive complex as
soon as it is formed. This would prevent further digestion of the
entire activation segment in case A and further activation in case B
so that it should be possible to isolate in case A the intact activation segment whereas in case B, a smaller peptide such as peptide a
should be recovered.

Pepstatin is an acylated pentapeptide containing two residues
of the unusual amino acid, 4-amino-3-hydroxy-6-methyl heptanoic acid
called statine (7) and is a tightly binding inhibitor of pepsin (8)
between pH 2 and 5. If it does not bind to pepsinogen then this
should serve as a useful trapping agent as long as intermediate
protein A is active and therefore capable of being inhibited. If
case B is correct, then intermediate protein A must be active or it
would not be possible for pepsin to form.

It has been suggested that glucose is covalently bound to porcine
pepsinogen but not pepsin (9) and, if this were so, the activation
process might be affected by this carbohydrate. Consequently, carbohydrates have been analyzed by gas-liquid chromatography on pepsinogens from different species.

MATERIALS AND METHODS

Pepsinogen from porcine stomachs was purified to homogeneity
(10) with a final step of chromatography on polylysine-Sepharose (11).
Polylysine-Sepharose was made from polylysine (12) obtained from
Cambrian Chemicals, Croydon, U. K. Bovine and canine pepsinogens,
chicken pepsinogen, calf prochymosin, and pepstatin were very

generous gifts from Professors Kassell, Kostka, Foltmann, and
Umezawa respectively. Rabbit muscle lactate dehydrogenase (LDH) and
deoxyribonuclease (DNase) were generously provided by Miles Research
Laboratories, Slough, U. K.; 1-ethyl-3-(3-dimethylamino propyl)carbo-
diimide, and pepsin were from Sigma (London) Ltd., U. K. Aminohexyl-
Sepharose was purchased from Pharmacia Fine Chemicals Ltd., Sweden.

Pepsinogen, prochymosin, DNase and LDH-5 concentrations were
determined by using values of extinction coefficient (A 280 nm of
1 mg/ml solutions) 1.47; 1.37; 1.15 and 1.49 respectively. Pepstatin
was dissolved in methanol and its concentration was determined by:
(a) amino acid analysis for alanine and valine after hydrolysis for
24 hr in vacuo with 6 M-HCl at 110°, and (b) titration against a
known amount of pepsin (8,13). Pepstatin was coupled through its
carboxyl group to aminohexyl-Sepharose using a water-soluble carbo-
diimide (14). Amino acid analysis and titration against standard
pepsin showed that the amount of pepstatin immobilized was 3.2 and
3.0 μmoles/g of dry resin, respectively.

Tests for Pepstatin Binding to Porcine Pepsinogen

Chromatography on this resin was carried out by applying the
protein in 0.1 M acetate buffer, pH 5.3/0.35 M NaCl or 0.1 M formate
buffer, pH 4.1/0.15 M NaCl. It was necessary to include salt to
prevent nonspecific adsorption. LDH-5 and DNase were used as con-
trols at pH 5.3 and 4.1 respectively to measure recoveries and to
assess nonspecific interaction with the pepstatin-aminohexyl-Sepharose. Chromatography of pepsinogen or LDH-5 at pH 5.3 on aminohexyl-
Sepharose (i.e., without pepstatin) showed that both proteins were
retarded nonspecifically until 0.35 M NaCl was included in the
buffer (14).

Binding of free pepstatin (in solution) to pepsinogen was tested
at several pH values (14) using a competitive assay technique devised
by Feinstein and Feeney (15).

Proteolytic activity measurements were made against hemoglobin
at pH 2 (16) and Azocoll at pH 5.3 (17).

Activation of Porcine Pepsinogen in the Presence of Pepstatin

Pepstatin was added to cold 3.2 mM-HCl, pH 2.5 followed by the
zymogen at 4°C (18). After incubation for 2.5 or 16 hr, the mixture
was freeze-dried and the resultant powder was extracted with 0.05 M
formate buffer pH 3.5. Most of the 'protein' was insoluble while the
supernatant contained the 'peptide'. The latter was applied to a
column of polylysine-Sepharose in formate buffer, pH 3.5. Variations
in these conditions were achieved by: (a) adding less pepstatin so

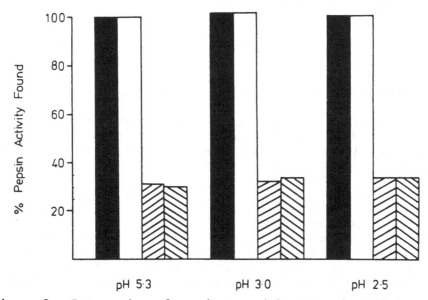

Figure 2. Interaction of pepsinogen with pepstatin. The
activity of pepsin (0.45 mg) against Azocoll at pH 5.3 was
measured in presence (▨▨▨) and absence (▬▬▬) of pepstatin (7 μg). Pepsinogen (0.25 mg) was preincubated, alone
(▭) or with pepstatin (▧▧▧) for 5 min at 0° in 0.05 M
acetate, pH 5.3 or 0.001 N-HCl or for 2 min at 0° in 0.0032
N-HCl. The pH was adjusted to 5.3 with a predetermined volume
of 0.2 N acetate, pH 5.3. Pepsin (0.45 mg) was added and the
mixtures were left for 1 min at 20°. Pepsin activity remaining
was measured against Azocoll.

that the pepstatin/pepsinogen molar ratio was 0.9:1, and (b) incubation in 0.05 M formate buffer, pH 4.1.

Carbohydrate Analysis of Porcine and Chicken Pepsinogens

Samples (2 mg) of the proteins were methanolyzed with methanol-
1.5 M HCl and N-acetylated with acetic anhydride (19). The determination of methyl glycosides as their trimethylsilyl ethers was carried
out by gas-liquid chromatography (20).

RESULTS

The Ability of Pepstatin to Bind to Porcine Pepsinogen

On chromatography of pepsinogen on immobilized pepstatin at
pH 5.3 and 4.1 (Table I), all of the applied pepsinogen was eluted
immediately from the affinity gel. Recoveries were comparable with
those of LDH and DNase neither of which would be expected to be retarded. At pH 3.0, however, a significantly lower yield of protein
(Fraction A-58%) was obtained and the 'missing' material was eluted
by 0.1 M NaHCO$_3$/1 M MaCl, pH 9 (Fraction B-28%). After concentration,
Fraction A was separated on polylysine-Sepharose at pH 6.5 into the
unretarded 'peptide' material (Fraction A1) and the 'protein' fraction
which had to be eluted with a salt gradient (14). The recovery of
peptide material in A1 (Table II) was 47% of what would have been
recovered if all of the pepsinogen applied to the pepstatin column
had activated itself.

Pepstatin immobilized through its carboxyl group should retain
most of its inhibitory capacity towards pepsin since various esters
of pepstatin have similar K_i values to that of native pepstatin
(8,21). Thus, pepsinogen seems to have little affinity for pepstatin
at pH 5.3 or 4.1 while at pH 3.0, during its passage down the column,
some of the zymogen apparently converted itself into pepsin and only
this was retarded by the column. Since 58% was recovered immediately,
the implication is that 40% of the zymogen activated itself to give
pepsin plus the activation segment peptides. Pepsin should have been
retarded by the resin and, indeed, Fraction B has the composition of
pepsin. The peptides should not have been retarded and should have
eluted with the unactivated pepsinogen. After concentration and
subsequent chromatography on polylysine-Sepharose, which binds pepsinogen and pepsin, but not peptides, the unretarded fraction A1 had
a composition fairly similar to that of the activation segment (14).
The ability of pepsinogen to interact with free pepstatin is shown in
Figure 2. The assay used (see legend) depends on measuring the free
inhibitor remaining after exposure to pepsinogen. At three pH values,

Table I

Chromatography of Pepsinogen on Pepstatin-Aminohexyl-Sepharose

Samples (1-2 mg in 0.1 ml) were applied to a column of pepstatin-aminohexyl-Sepharose (0.5 cm x 2.2 cm) at $4^{\circ}C$ in the
appropriate buffer. Elution was continued with the same buffer for
3 bed-volumes, when the eluent was changed to 0.1 M-NaHCO$_3$/1M-NaCl,
pH 9. Recoveries are calculated from protein absorbance at 280 nm.

Protein	pH at which applied	% Recovery on elution at		No. of runs
		Initial pH	pH 9	
Pepsinogen	pH 5.3[a] +0.35M NaCl	83	0	4
Pepsin	pH 5.3[a] +0.35M NaCl	0	82	4
Lactate dehydrogenase	pH 5.3[a] +0.35M NaCl	83	0	3
Pepsinogen	pH 4.1[b] +0.15M NaCl	87	0	4
Pepsin	pH 4.1[b] +0.15M NaCl	0	84	4
Deoxyribonuclease	pH 4.1[b] +0.15M NaCl	85	0	3
Pepsinogen	pH 3.0[b] + 1M NaCl	58	28	4

[a] In 0.1M acetate buffer
[b] In 0.1M formate buffer

Table II

Amino Acid Compositions of Fractions Obtained by Chromatography
of Pepsinogen on Pepstatin-Aminohexyl-Sepharose at pH 3.0

Samples were hydrolyzed for 20h in vacuo in 6M HCl at 105°C.
Values are not corrected for losses on hydrolysis. N.D., Not
determined.

Amino acid	Fraction A1		Fraction B	
	Found	Expected from activation segment[a]	Found	Expected from pepsin
		Residues/molecule		Residues/molecule
Lys	4.60	9	1.21	1
His	1.34	2	1.20	1
Arg	1.14	2	1.94	2
Asp	4.00	4	42.52	42
Thr	1.28	1	24.75	26
Ser	2.16	2	40.36	43
Glu	2.52	3	26.45	26
Pro	2.26	3	N.D.	16
Gly	2.20	1	37.87	35
Ala	2.88	4	16.35	16
Val	2.20	3	17.82	22
Met	0.00	0	1.97	4
Ile	1.32	1	20.11	26
Leu	4.85	7	27.29	26
Tyr	0.95	1	16.36	16
Phe	1.36	2	14.00	14

[a]The activation segment of pepsinogen contains residues 1-44
as shown in Figure 1.

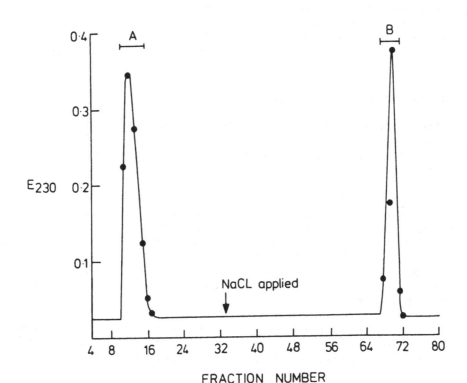

Figure 3.   Chromatography of "supernatant" on polylysine-
Sepharose.   The column was 1.8 x 16 cm equilibrated in 0.05 M
formate, pH 3.5.   After washing with 70 ml of buffer, a linear
gradient of 70 ml buffer plus 70 ml 1 M NaCl in buffer was
applied.   Fractions (2 ml) were collected.

none of the pepstatin appeared to have been removed by complexing with pepsinogen so that it was all available to inhibit the pepsin added. An excess of pepsin over pepstatin was used deliberately so that there was always a positive value of pepsin activity to be measured. Control incubations showed there was no activation of pepsinogen under the conditions used nor was there any activation of pepsinogen by pepsin at pH 5.3. Porcine pepsinogen undergoes a conformational change on exposure to low pH where it autoactivates (2,22,23) even in the cold; so it is impossible to measure binding of pepstatin directly at this pH, but qualitatively at least, pepstatin apparently has little affinity for pepsinogen compared to pepsin. This is confirmed by the immobilized pepstatin studies.

Activation of Porcine Pepsinogen in the Presence of Pepstatin

After activation at pH 2.5 in the presence of a 1.4 fold molar excess of pepstatin, the incubation mixture was freeze-dried and extracted with formate buffer giving an insoluble 'protein' and a supernatant fraction.

Insoluble material. This was completely inactive against hemoglobin at pH 2 and Azocoll at pH 5.3. NH_2-terminal analysis with dansyl chloride showed both isoleucine and leucine; gel electrophoresis at pH 8.1 revealed two bands, one with a mobility similar to that of pepsinogen and the other major band with a mobility intermediate to those of pepsinogen and pepsin. The composition (Table III) is similar to that calculated from the sequence of residues 17-370 in porcine pepsinogen.

When the pepstatin/pepsinogen molar ratio was lowered to 0.9/1, this insoluble 'protein' fraction had 10% of the specific activity of pepsin against hemoglobin and Azocoll (Table III).

Supernatant. This was chromatographed on polylysine-Sepharose at pH 3.5 to separate the unretarded 'peptide' fraction (A) from the 'protein' fraction (B), which was very similar to the insoluble material already described, and which was eluted with a linear NaCl gradient (Fig.3).

The composition of the 'crude' fraction A (Table IV) was very similar to that calculated from the sequence of residues 1-16 in porcine pepsinogen; it made little difference to the results whether the activation was above (inter) or below (intramolecular) pH 3. The compositions show alanine and valine from the excess of pepstatin added, since this compound is not retarded by polylysine-Sepharose. It has an acylated NH_2-terminus and so is not reactive to dansyl chloride. The yield of 'crude' peptide averaged about 60%.

Table III

Amino Acid Compositions of the Insoluble 'Protein' Fractions

Samples were hydrolyzed for 24 hr in vacuo in 6M HCl at 105°C. Values are not corrected for losses on hydrolysis.

Residues/molecule

Amino acid	'Protein' from pepstatin + pepsinogen (molar ratio 1.4:1)	Expected from sequence of residues 17-370	'Protein' from pepstatin + pepsinogen (molar ratio 0.9:1)	Pepsin (residues 45-370)
Lys	8.05	7	1.37	1
His	2.96	3	1.07	1
Arg	3.29	2	1.88	2
Asp	42.3	45	44.6	42
Thr	23.0	27	25.2	26
Ser	37.2	44	40.0	43
Glu	27.9	27	28.3	26
Pro	16.7	17	17.1	15
Gly	35.7	36	39.1	35
Ala	22.0	20+1[a]	20.7	17+1[a]
Val	23.0	22+2[a]	21.5	22+2[a]
Met	3.4	4	3.7	4
Ile	24.3	27	22.6	26
Leu	32.4	29	27.7	26
Tyr	15.5	17	15.0	16
Phe	16.0	16	14.0	14

[a] Including 1 and 2 mol of alanine and valine from 1 mol of pepstatin bound/mol of pepsin.

Table IV

Amino Acid Compositions of the 'Peptide' Fractions

Samples were hydrolyzed for 24 hr in vacuo in 6 M HCl at 105°C. Values are not corrected for losses on hydrolysis. Data are in residues per molecule.

| Amino acid | Pepstatin/pepsinogen molar ratio 1.4:1 | | Expected from sequence of residues 1-16 | Pepstatin/pepsinogen molar ratio 0.9:1 | Expected from sequence residues 1-44 |
	Activation at pH 2.5[a]	Activation at pH 4.1[c]		Activation at pH 2.5[c]	
Lys	2.68	2.18	3	6.60	9
His	0	0	0	1.76	2
Arg	1.62	1.45	2	1.45	2
Asp	1.00	1.00	1	3.97	4
Thr	0	0	0	1.45	1
Ser	1.03	1.17	1	2.37	2
Glu	1.12	1.24	1	2.43	2
Pro	0.95	0.99	1	3.14	3
Gly	0.73[b]	1.13[b]	0	1.91	1
Ala	1.03[b]	1.45[b]	0	2.82	4
Val	3.83[b]	4.64[b]	3	2.33	3
Ile	0	0	0	1.06	1
Leu	3.43	3.62	4	5.10	7
Tyr	0	0	0	1.15	1
Phe	0	0	0	2.60	2
NH2-terminus	Leu	Leu	Leu	Many	

[a]Mean values from the eight activations performed.

[b]Including 1 and 2 mol of residues of alanine and valine/mol from excess of pepstatin.

[c]One activation experiment.

The excess of pepstatin (peak A2) was removed on a column of Bio-Gel P2 (Fig.4). The composition and terminal analyses of Al are given in Table V. Carboxyl-terminal residue was determined from a time-course of digestion with carboxypeptidase A. High-voltage paper electrophoresis at pH 6.5 and thin-layered chromatography showed Al to contain one major basic peptide but, in both cases, very faint additional spots could be detected.

When the molar ratio of pepstatin/pepsinogen was 0.9:1, the composition of the 'crude' peptides obtained was much more similar to the expected composition from residues 1-44 (Table IV). Several NH_2-termini were obtained similar to the result obtained on activation of pepsinogen in the absence of pepstatin. Presumably the pepstatin inhibits 90% of the activating pepsinogen but the remaining 10% can convert itself all the way into pepsin. This active pepsin could then split off the remaining activation segments exposed on the inhibited molecules so that a mix of 90% inhibited pepsin plus 10% active pepsin is obtained.

Inhibition of Pepsin by Peptide Al

It has long been known (24) that one of the peptides released on activation of pepsinogen could act as a pepsin inhibitor at pH values above 4. Peptide 1-16 from porcine pepsinogen (25) and the corresponding peptide 1-17 from bovine pepsinogen (5) can indeed inhibit the milk-clotting activity of pepsin above pH 5 (see Kumar et al., Chapter 13 this volume).

Since peptide Al appeared to have been derived from residues 1-16, it was of interest to see whether this peptide, generated in the course of a 'stopped' activation, could inhibit pepsin and the homologous milk-clotting calf protein, chymosin. The results (V. Barkholt Pedersen and P. H. Ward - personal communications) show that Al had no effect on the milk-clotting activity of chymosin (confirming that all of the pepstatin must have been removed) while independently the Danish and American labs produced similar curves for the inactivation of porcine pepsin (Fig.5). Thus, it would seem that the first peptide produced on activation is the inhibitor peptide. This seems to make sense from a control point of view, i.e., the first bond being hydrolyzed on activation releases the pepsin inhibitor so it can exert its effect if required.

In view of the sequence homologies between pepsinogen and prochymosin (Fig.1) it was a little suprising that Al could not inhibit chymosin. This suggested that prochymosin might activate itself by a different mechanism that never generates a 16-residue peptide.

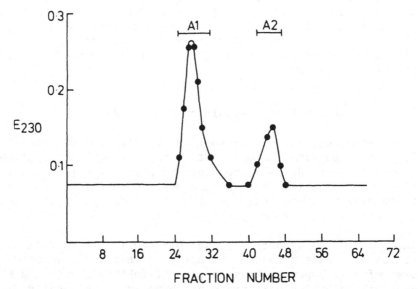

Figure 4.   Chromatography of "crude" peptide on Bio-Gel P2.
The column was 1.3 x 100 cm, equilibrated in 0.5% ammonium
bicarbonate.   Fractions (2 ml) were collected.

Table V

Amino Acid Composition of Peptide A1

Samples were hydrolyzed for 72 hr in vacuo in 6N HCl at 105°

Amino acid	Found in A1 experimentally	Expected from residues 1-16
	Residues/molecule	Residues/molecule
Lys	2.86	3
Arg	1.90	2
Asp	1.00	1
Ser	1.00	1
Glu	1.07	1
Pro	0.89	1
Gly	0.16	0
Val	2.84	3
Leu	3.94	4
NH_2-Terminus	Leu,Bis-Lys(Faint) Val (Very Faint)	Leu
COOH-Terminus	Leu	Leu

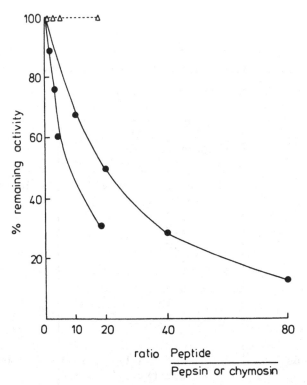

Figure 5. The effect of peptide A1 on the milk-clotting activities of pepsin (—●——●—) and chymosin (···▲····▲···) at pH 5.3. The peptide was premixed with the enzyme before measuring the clotting activity remaining.

Activation of Bovine, Canine, and Chicken Pepsinogens and Calf Prochymosin

Samples of these proteins were subjected to activation and conditions (with pepstatin at a molar ratio of 1.4 to 1) similar to those described for porcine pepsinogen.

The supernatants from bovine and canine pepsinogens were fractionated on polylysine-Sepharose at pH 3.5 as already described whereas, since neither prochymosin nor chicken pepsinogen has affinity for this resin, these preparations were chromatographed on DEAE-cellulose in 0.02 M phosphate buffer, pH 5.4, which separated the unretarded 'peptide' fraction from the adsorbed 'protein'.

The bovine 'peptide' has a similar composition (Table VI) to the NH_2-terminal 17 residues in bovine pepsinogen (Fig.1) which are homologous with the sequence already discussed for porcine pepsinogen. In contrast, the 'peptide' fraction from calf prochymosin has a composition (Table VI) resembling that of residues 1-27 in the prochymosin sequence (Fig.1). After removal of the excess of pepstatin on Bio-Gel P2, the prochymosin 'peptide' had the composition shown in the last column of Table VI. For the sake of brevity the details of the 'protein' fraction from prochymosin are not given but this showed a composition intermediate to prochymosin and chymosin, particularly in the values for lysine and leucine which reflect the largest changes.

The 'peptide' fractions from activation of canine and chicken pepsinogens are shown in Table VII. The composition of canine pepsinogen is known (26) but that of the pepsin is not, so it is not possible to calculate a 'difference' for the activation segment. Assuming a modest homology with the proteins in Figure 1, however, it seems reasonable to assume that about 50 residues should be released. Only about 22 amino acids were released in the presence of pepstatin.

The compositions of chicken pepsinogen (27) and pepsin (27,28) are known and the 'difference' calculated for the activation segment is given in Table VII Activation in the absence of pepstatin gave a 'peptide' composition (Table VII) in reasonable agreement with these predicted values. In contrast, inclusion of pepstatin during the activation produced significantly smaller numbers of residues.

The composition of the protein generated on activation of chicken pepsinogen in the absence of pepstatin (Table VIII) bears a marked resemblance to chicken pepsin (the low Met and Tyr values were due to poor hydrolysis conditions) whereas the protein obtained in the presence of pepstatin has a composition intermediate to those of the zymogen and pepsin and quite similar to the 'calculated' values obtained by subtraction of the peptide composition (Table VII) from that of pepsinogen.

Table VI

Amino Acid Compositions (mol of residue/mol) of the 'Peptide' Fractions
from Bovine Pepsinogen and Calf Prochymosin

Samples were hydrolyzed in vacuo for 24 hr in 6N HCl at 105°.

Amino acid	Peptide from bovine pepsinogen	Expected from sequence of residues 1-17	Peptide from calf prochymosin	Expected from sequence of residues 1-27	Prochymosin peptide after Bio-gel P2
Lys	3.14	4	2.75	4	3.10
His	0	0	0.78	1	1.06
Arg	1.00	1	1.63	2	1.66
Asp	1.42	1	1.38	1	1.38
Thr	0	0	1.10	1	1.00
Ser	2.70	2	1.58	1	1.12
Glu	2.22	1	2.92	3	2.94
Pro	0.68	1	0.81	1	1.08
Gly	1.83	0	2.15	2	2.25
Ala	2.06[a]	0	1.93[a]	2	1.74
Val	4.66[a]	3	1.31[a]	0	0.30
Met	0	0	0	0	0
Ile	0.93	1	1.82	2	1.65
Leu	2.45	3	4.01	5	4.71
Tyr	0	0	1.11	1	0.75
Phe	0	0	1.00	1	1.00
Yield of Peptide	50%		80%		

[a] Including 1 and 2 mol of residues of alanine and valine/mol from excess of pepstatin.

Table VII

Amino Acid Compositions (mol of residue/mol) of the 'Peptide' Fractions
from Canine and Chicken Pepsinogens
Samples were hydrolyzed in vacuo for 48 hr in 6M HCl in 105°

Amino acid	Peptide(s) from canine pepsinogen	Residues released on activation	Chicken pepsinogen minus pepsin(27)	Residues released on activation in absence of pepstatin	Peptide(s) from chicken pepsinogen in presence of pepstatin
Lys	1.76		10	10.00	3.07
His	0.92		5	4.80	1.34
Arg	0.88		3	4.80	1.32
Asp	3.14		8	5.80	2.31
Thr	0		4	2.90	0.94
Ser	1.08		4	N.D.	1.71
Glu	2.00		7	N.D.	2.04
Pro	0.94	(NOT KNOWN)	5	4.54	1.48
Gly	1.44		5	5.52	2.06
Ala	1.22[a]		3	4.82	1.48[a]
Val	2.02[a]		5	5.43	1.99[a]
Met	0		0	0.86	0.07
Ile	1.62		3	3.84	1.09
Leu	3.86		10	9.93	3.10
Tyr	0		4	3.70	0.90
Phe	1.04		3	2.83	1.00
Total	approx. 21	approx. 50	79	72+	approx. 25
Recovery	35%	-	-	100%	65%

[a] Including 1 and 2 mol of residues of alanine and valine/mol from excess of pepstatin.

Table VIII

Amino Acid Compositions (mol of residue/mol) of the 'Protein' Fractions from Activation of Chicken Pepsinogen in the Absence and Presence of Pepstatin

Samples were hydrolyzed in vacuo for 24 hr in 6M HCl at 105°

Amino acid	Protein from activation without pepstatin	Chicken pepsin (27)	Chicken pepsin (28)	Chicken Pepsinogen (27)	Protein from activation with pepstatin	Calculated values for pepsinogen minus 'peptide'
Lys	8.00	8	8	18	15.00	15.0
His	3.64	3	4	8	6.6	6.4
Arg	4.95	4	4	7	7.4	7.8
Asp	34.72	35	32	43	38.8	41.4
Thr	20.91	24	22	28	23.7	26.9
Ser	29.63	35	38	39	31.6	35.7
Glu	23.21	23	20	30	31.4	N.D.
Pro	11.59	14	12	19	17.9	17.0
Gly	30.60	27	26	32	38.2	30.0
Ala	16.68	15	14	18	22.2[a]	16.5
Val	22.42	21	17	26	28.4[a]	24.1
Met	4.93	9	7	9	10.1	9.4
Ile	18.41	20	14	23	25.9	21.5
Leu	20.49	20	18	30	30.2	27.4
Tyr	15.56	20	16	24	25.5	19.2
Phe	17.18	18	18	21	25.1	18.5

[a] Including 1 and 2 mol of residues of alanine and valine/mol from excess of pepstatin.

Thus, in the cases of bovine pepsinogen and calf prochymosin where the activation segment sequences are known, peptides were re-leased corresponding to the NH_2-terminal parts of the sequence only. For the canine and chicken proteins where no sequence data is availa-ble, it appears that inclusion of pepstatin stops the activations before all of the activation segment is released. Consequently, a sequential activation mechanism and not a one-step transformation must be operative with all five of these zymogens.

Carbohydrate Analysis on Porcine and Chicken Pepsinogens

It has been suggested (9) that porcine pepsinogen contained covalently bound carbohydrate while pepsin did not. If so, this might provide a convenient label for following the intermediate states of activation. It has also been suggested (27) that chicken pepsinogen is a glycoprotein. Analyses of these two proteins by a sensitive gas-liquid chromatography detection system was carried out by Professor J. R. Clamp (in contrast to the previously used colorimetric determi-nations). Porcine pepsinogen prepared by conventional methods (10) with a final step of chromatography on polylysine-Sepharose contains no carbohydrate whatever (Table IX repeated with three different pre-parations). Chicken pepsinogen on the other hand contains mannose and N-acetyl-glucosamine. These two residues are commonly found in glyco-proteins often with the N-acetyl glucosamine linked to an asparagine side chain in the polypeptide. The values obtained are in excellent agreement with the 6-7 'glucose' equivalents found previously (27). Thus, chicken pepsinogen is probably a glycoprotein whereas porcine pepsinogen is not.

DISCUSSION

It is suggested that the process being investigated was the intra-molecular, self-catalyzed activation of pepsinogen, since: (A) the pepsinogen samples used in this study were all prepared in such a way that they could not contain active pepsin (denatured at pH 7 and above; (B) activation was carried out at pH 2.5 (and 4.1, and the same result was obtained for the intermolecular catalyzed reaction); and (C) no active protein could ever be detected, except when the molar ratio of pepstatin/pepsinogen was less than 1:1.

Porcine pepsinogen apparently does not convert itself into pepsin in a one-step transformation, releasing the activation segment 1-44 intact. By difference, it would seem that the alternative, progres-sive activation scheme must be operative. Certainly, the amino acid composition, and the NH_2- and COOH-terminal analyses on the peptide obtained from the 'stopped' activation plus the fact that it acts like the pepsin inhibitor peptide all suggest it to have been derived from residues 1-16 in the porcine sequence.

Table IX

Carbohydrate Compositions (mol of residue/mol) of Porcine and
Chicken Pepsinogens

Samples were hydrolyzed in methanolic-HCl for 24 hrs at 90° (20)

Sugar	Porcine pepsinogen	Chicken pepsinogen
Fucose	0	0.04
Mannose	0	2.24
Galactose	0	0.28
Glucose	0	0.12
N-acetyl-glucosamine	0	4.48
No. of analyses	3	1

The use of pepstatin to stop the activation, of course, precludes identification of the subsequent steps in the transformation.
It is uncertain whether a limited number of discrete steps are involved or whether a ragged activation at many susceptible bonds
could occur. The results with pepsinogens from the other species
tested support the idea that activation proceeds through a limited
number of intermediate proteins; e.g., with bovine pepsinogen pepsinogen whose activation segment sequence is known (5) a similar peptide to the porcine cases was obtained.

Calf prochymosin produced a slightly different results in that a
larger peptide of 27 residues was released before avid binding of pepstatin stopped the proteolysis. This implies that prochymosin does
not undergo its initial activating split at the Leu 16-Lys 17 bond
(homologous with Leu 16-Ile 17 in porcine and bovine pepsinogens-
porcine numbering) but at the Phe 25-Leu 26 bond (see Fig.1). Very
recently (V. Barkholt Pedersen - personal communication) it has been
shown that prochymosin covalently attached to Sepharose undergoes a
monomolecular activation in which the initial cleavage, as determined
by sequence analysis, takes place between Phe-25 and Leu-26. A similar study with insolubilized bovine pepsinogen suggested that the
initial cleavage was at the bond between Leu-16 and Ile-17. Thus, it
would appear that while prochymosin activates itself by a sequential
mechanism, like pepsinogen, the actual nature of the intermediates
involved in the two processes are different.

Since the initial experiments with pepstatin were carried out
(30), Tang and coworkers have tried a similar approach at much lower
protein concentrations and at pH 2 (31). Some carefully performed
kinetic studies with globin binding (see (31) and J. A. Hartsuck
et al., Chapter 6 in this volume) have confirmed that the conversion
of pepsinogen to pepsin is not a one-step transformation; they have
shown the existence of an intermediate, δ, which can bind globin reversibly and whose disappearance constitutes the rate limiting step.

Their implication is that this intermediate has simply undergone
conformational alterations and not limited proteolysis as we have
found. As yet details are not available for the structure of this
intermediate nor is there any evidence for other intermediates beyond
this state, only that this intermediate undergoes a further rate-
limiting reaction. Our interpretation of limited proteolysis to
generate intermediate protein A would fit these kinetic data equally
well, so that resolution of this problem awaits further investigation.
Nevertheless, Tang and coworkers have some evidence to suggest that
pepstatin can bind to a form of pepsinogen at pH 2, and with our
activation experiments in the presence of pepstatin, the yield of peptide 1-16 obtained was seldom greater than 60% (this was not due to
mechanical losses). Thus, it may be that under some conditions, pepstatin can bind to pepsinogen, but irrespective of this, since proteolysis is an essentially irreversible event and since the only peptide

ever observed as derived from the NH_2-terminal 16 residues and never
the entire 1-44 segment, the possibility of a one-step activation
mechanism must be excluded.

ACKNOWLEDGEMENTS

This work was supported by the M.R.C. (Grant No. G974/126/B).
We are very grateful to Professor J. R. Clamp, University of Bristol,
for carrying out the carbohydrate analyses.

SUMMARY

By incubation of porcine, bovine, canine, or chicken pepsinogens
and calf prochymosin with pepstatin at pH 2.5, the first active protein generated on activation is trapped in an inactive complex. The
first activation peptide liberated from porcine pepsinogen has been
identified as residues 1-16 whereas that from prochymosin is derived
from residues 1-27. This suggests that pepsin and chymosin are not
formed by one-step conversions from their zymogens, but by (different)
sequential, activation mechanisms.

REFERENCES

1.   Al-Janabi, J., Hartsuck, J. A., and Tang, J. (1972) J. Biol.
     Chem. 247, 4628-4632
2.   McPhie, P. (1972) J. Biol. Chem. 247, 4277-4281
3.   Sanny, C. G., Hartsuck, J. A., and Tang, J. (1975) J. Biol.
     Chem. 250, 2635-2639
4.   Sepulveda, P., Marciniszyn, J. P., Liu, D., and Tang, J. (1975)
     J. Biol. Chem. 250, 5082-5088
5.   Harboe, M., Andersen, P. M., Foltmann, B., Kay, J., and Kassell,
     B. (1974) J. Biol. Chem. 249, 4487-4494
6.   Pedersen, V. B., and Foltmann, B. (1975) Eur. J. Biochem. 55,
     95-103
7.   Tang, J. (1976) Trends in Biochem. Sci. 1, 205-208
8.   Aoyagi, T., Kunimoto, S., Morishima, H., Takeuchi, T., and
     Umezawa, H. (1971) J. Antibiot. 24, 687-694
9.   Neumann, H., Zehavi, U., and Tanksley, T. D. (1969) Biochem.
     Biophys. Res. Commun. 36, 151-155
10. Ryle, A. P. (1970) Methods Enzymol. 19, 316-336
11. Kay, J. (1972) Abstr. Fed. Eur. Biochem. Soc. Meet. 8th,
     Amsterdam, Abstr. No. 458
12. Nevaldine, B., and Kassell, B. (1971) Biochim. Biophys. Acta
     250, 207-209
13. Kunimoto, S., Aoyagi, T., Nishizawa, R., Komai, T., Takeuchi,
     T, and Umezawa, H. (1974) J. Antibiot. 27, 413-418

14. Kay, J., and Dykes, C. W. (1976) Biochem. J. 157, 499-502
15. Feinstein, G., and Feeney, R. E. (1966) J. Biol. Chem. 241, 5183-5189
16. Anson, M. L. (1948) in Crystalline Enzymes (Northrop, J. H., Kunitz, M., and Herriott, R. M. eds) pp. 305, Columbia Univ. Press, New York.
17. Kay, J. (1975) Anal. Biochem. 67, 585-589
18. Dykes, C. W., and Kay, J. (1976) Biochem. J. 153, 141-144
19. Clamp, J. R., Dawson, G., and Hough, L. (1967) Biochim. Biophys. Acta 148, 342-349
20. Bhatti, T., Chambers, R. E., and Clamp, J. R. (1970) Biochim. Biophys. Acta 222, 339-347
21. Knight, C. G., and Barrett, A. J. (1976) Biochem. J. 155, 117-125
22. Funatsu, M., Harada, Y., Hayashi, K., and Jirgensons, B. (1971) Agr. Biol. Chem. 35, 566-572
23. Wang, J. L., and Edelman, G. M. (1971) J. Biol. Chem. 246, 1185-1191
24. Herriott, R. M. (1941) J. Gen. Physiol. 24, 325-338
25. Anderson, W., and Harthill, J. E. (1973) Nature, 243, 417-419
26. Marciniszyn, J. P., and Kassell, B. (1971) J. Biol. Chem. 246, 6560-6565
27. Bohak, Z. (1969) J. Biol. Chem. 244, 4638-4648
28. Green, M. L., and Llewellin, J. (1973) Biochem. J. 133, 105-115
29. Pedersen, V. B. (1976) Abstr. X[th] Int. Union Biochem. Meet. Hamburg, Abstr. No. 04-3-319
30. Kay, J., and Dykes, C. W. (1975) Abstr. Fed. Eur. Biochem. Soc. Meet. X[th], Paris, Abstr. No. 758
31. Marciniszyn, J. P., Huang, J. S., Hartsuck, J. A., and Tang, J. (1976) J. Biol. Chem. 251, 7095-7102

CATALYTIC MECHANISM OF PEPSIN

SPECIFICITY AND MECHANISM OF PEPSIN ACTION

ON SYNTHETIC SUBSTRATES

Joseph S. Fruton

Kline Biology Tower, Yale University
New Haven, Connecticut

Studies on the specificity and mechanism of pepsin action have
involved the use of several types of synthetic substrates. In the
early work, substrates related to Z-Glu-Tyr (1) or Ac-Phe-Tyr (2)
were largely employed; for a review, see (3). In particular, much
use had been made of acyl dipeptides of the type A-X-Y, where X and
Y are aromatic L-amino acid residues forming the sensitive peptide
bond. Because the pK_a of the carboxyl group falls in the pH range of
pepsin activity, the pH dependence of the kinetic parameters is a
function of the ionization of prototrophic groups in both the enzyme
and the substrate. The methyl or ethyl esters, or amides, of acyl
dipeptides have also been used extensively, but because of their
limited solubility in aqueous solution, variable amounts of organic
solvents had to be added. Such solvents, even in relatively low concentration, markedly inhibit pepsin action (4,5). To obviate this
difficulty, another type of pepsin substrate was introduced, in which
a cationic group (the imidazolium group of a His residue, the α-
ammonium group, a pyridinium group, or a morpholinium group) is
present. By means of such substrates, in particular of the general
structure Z-His-X-Y-OMe, the primary specificity of pepsin was defined as a preference for hydrophobic L-amino acid residues in both
the X- and Y- positions (10); substitution of either Phe residue of
Z-His-Phe-Phe-OMe by its D-enatiomer renders the X-Y bond resistant
to pepsin action (6). The favorable effect of an aromatic and planar
substituent at the β-carbon of the X and Y residues was emphasized
by the finding that when X is β-cyclohexyl-L-alanyl, the value of
k_{cat} is much lower than that found for the corresponding substrate
in which X or Y = Phe, and is similar to that for substrates in which
the X- or Y-position is occupied by an aliphatic amino acid residue
larger than Ala (Nva, Nle, Leu, Met). Apparently, the side chains of

these amino acids can interact with a portion of the enzymic region
that binds planar aromatic groups. It was also shown that the replacement of Phe in the X-position by Val or Ile rendered the X-Y
bond resistant to pepsin action than when X = Gly, indicating that
when the X-position is occupied by a residue that is branched at
the β-carbon, one of the catalytic groups of pepsin may be prevented
from attacking the carbonyl group of the sensitive bond. Moreover,
the importance of the β-methylene group as a structural element of
the X-Y unit was underlined by the finding that replacement of
either Phe residue of Z-His-Phe-Phe-OMe by a L-Phenylglycyl residue
also rendered the X-Y bond resistant to pepsin action (11).

The available data on the primary specificity of pepsin may be
summarized in terms of a strong preference for two aromatic L-amino
acid residues forming the sensitive X-Y bond, the best substrates
being those in which X= Phe and Y = Trp, Tyr, or Phe. The introduction of a Phe(4NO$_2$) residue in the X-position did not alter
markedly the kinetic parameters of pepsin action; this permitted
the development of a spectrophotometric method for following the
hydrolysis of the Phe(NO$_2$)-Phe bond (6). In contrast to the widely
used analytical procedures for estimating the rate of formation of
the amine product (e.g., Phe-OMe) by means of its reaction with
ninhydrin or fluorescamine, this method measures the rate of formation of the acidic product e.g., Z-His-Phe(4NO$_2$).

In addition to the nature of the amino acid chains and the con-
figuration of the dipeptidyl unit providing the sensitive bond, the
primary specificity of pepsin appears to involve a requirement for
both the N-terminal NH group and the C-terminal CO group of the dipeptidyl unit. This inference is drawn from the finding that
O-acetyl-β-phenyl-L-lactyl-L-Phenylalanine (12) and Z-His-Phe-L-
phenylalaninol (6) are resistant to pepsin action under conditions
where Ac-Phe-Phe and Z-His-Phe-Phe-OMe are readily cleaved by the
enzyme.

Studies on the modification of the A and B groups of cationic
pepsin substrates of the type A-Phe-Phe-B led to the recognition
that secondary enzyme-substrate interactions, at a distance from
the site of catalytic action, markedly affect k_{cat}, without comparable changes in the value of K_m; for a recent review, see (13). This
result was interpreted to indicate flexibility in an extended active
site (7,13). From the effect of elongation of substrates of the
type A-Phe-Phe-B or A-Phe(4NO$_2$)-Phe-B at the amino and carboxyl
termini of the sensitive Phe-Phe unit, it was inferred that the
active site of pepsin can accomodate at least a heptapeptide; in a
fully extended conformation, this would correspond to a distance of
about 25 A (14). We have eschewed the designation of the segments
of the extended active site that are complementary to individual
amino acid residues as "subsites", in the manner proposed by
Schechter and Berger (15), because such designation appears to be

inappropriate to a structure that seems to undergo significant change upon enzyme-substrate interaction (6,16).

Data for cationic substrates of the type A-Phe-Phe-OP4P, where A is a Z-dipeptidyl group, indicate that this group interacts with the enzyme as a unit and that the benzyl portion of the Z group participates significantly in the interaction (17). To study more directly the secondary interaction of such an amino-terminal substituent with pepsin, the Z group was replaced by a mansyl or dansyl group, which is relatively weakly fluorescent in aqueous solution, but which fluoresces strongly in a less polar environment, such as that in the active-site groove of pepsin (18,19).

It is important to note that the active site of pepsin has relatively little intrinsic affinity for the fluorescent probe group, when it is present in compounds such as mansylamide or Mns-Gly-Gly-OP4P (19,20). When, however, the mansyl group is attached to a peptide having a Phe-Phe unit strongly bound at the catalytic site of the enzyme (21,22), the fluorescence of the probe is greatly enhanced. The mansyl group has thus been drawn into the extended active site of pepsin by virtue of the primary specificity of the enzyme toward the Phe-Phe unit of the substrate. With Mns-Phe-Phe-OP4P, the enhanced fluorescence of the mansyl group is reduced by the addition of pepstatin to the small fluorescence intensity observed with Mns-Gly-Gly-OP4P. The fluorescence of the complex of pepsin with the latter mansyl compound is not altered by the addition of pepstatin (19) and this result is consistent with earlier data (23) indicating that pepsin has a weak binding locus for the mansyl group, distinct from the extended active site of the enzyme.

Stopped-flow kinetic measurements under conditions of enzyme excess have shown that a rapid increase in mansyl fluorescence on mixing a substrate with pepsin is followed by a first-order decrease in fluorescence when the affinity of the acidic cleavage product for the active site is much less than that of the substrate. Such kinetic measurements have given estimates of K_s, the dissociation constant of the rate-limiting Michaelis complex, and in all cases $K_s = K_m$, the Michaelis constant estimated under conditions of substrate excess by determination of the rate of formation of the amine product (24). This result strengthens the validity of earlier conclusions (3,13) that the rate-limiting step in the action of pepsin on peptide substrates under conditions of substrate excess is the decomposition of the first detectable enzyme-substrate complex, and that no kinetically significant intermediate accumulates in the process.

For most of the substrates tested, the value of k_{cat} estimated from steady-state kinetic measurements under conditions of substrate excess was equal (within the precision of the data) to the value of k_2, the first-order rate constant for the decomposition of the ES complex, determined under conditions of enzyme excess (24).

With the most sensitive substrates (e.g., Dns-Ala-Ala-Phe-Phe-OP4P), however, k_2 was found to be much greater than k_{cat}. Further work is needed to explain this result.

Whereas the acyl dipeptide substrates of pepsin were resistant to acid proteinases such as cathepsin D (25,26), several of the more sensitive cationic substrates were cleaved by these and other members of this family of proteinases (27,28). The apparent differences in the specificity of the action of these enzymes have been shown to be a consequence, in large part, of the contribution made by secondary enzyme-substrate interactions to catalysis (29).

In addition to the types of synthetic substrate for pepsin mentioned above, others include depsipeptides such as Z-His-Phe(4NO2)-Pla-OMe, whose rapid hydrolysis showed that pepsin can act as an esterase provided the primary specificity requirements are met (30), sulfite esters such as bis-p-nitrophenyl sulfite (31,32) which are also cleaved rapidly, and various more resistant compounds such as Tfa-Phe (33) and Leu-Tyr-Leu or Leu-Tyr-NH2 (34), whose mode of cleavage by pepsin has also suggested various hypotheses about the mechanism of pepsin action.

In earlier considerations of the specificity and mechanisms of proteolytic enzymes in their action of synthetic substrates, it has been customary to separate binding specificity as measured by K_S (or by K_m when it has been shown to approximate K_S) from kinetic specificity as measured by k_{cat} (35). The recent results obtained with pepsin, as well as those for other proteinases, notably papain (16) and elastase (36) suggest that this separation is inappropriate. With the recognition that the extended active sites of many protein-ases interact specifically with oligomeric substrates at loci distant from the site of catalytic action, and that such secondary inter-actions can markedly affect k_{cat} without significant change in K_S, binding specificity and kinetic specificity have been increasingly considered to be inextricably linked (13). Attention has been drawn to the possibility that a significant portion of the binding energy in enzyme-substrate interaction may be used to decrease the free energy of activation in the bond-breaking process (37). One of the molecular mechanisms whereby the transformation of binding energy into kinetic efficiency might be effected is through conformational changes at the catalytic site induced by the multiple cooperative interaction of complementary portions of the extended active site of a proteinase with an oligomeric substrate. Such interaction may be expected to be a stepwise process (38), and in the case of the bind-ing to papain of suitable fluorescent substrates (e.g., Mns-Gly-Val-Glu-Leu-Gly) it has been shown by stopped-flow measurements that a very bimolecular association reaction is followed by at least one much slower first-order process leading to the Michaelis complex in which the Glu-Leu bond of the substrate is cleaved (39). In the

stopped-flow experiments on the interaction of pepsin with substrates
such as Mns-Gly-Phe-Phe-OP4P, it should be noted that such a two-step
process was not seen. It is a reasonably hypothesis that the conformational changes in the pepsin molecule are much more rapid than
in the case of papain. Indeed, in the family of the proteinase,
there is likely to be a continuum of flexibility, with trypsin,
which is lease responsive to secondary interactions (29), representing the more rigid active site, pepsin the more flexible active site
and papain intermediate between the two. In a multi-step interaction
between a proteinase and an oligomeric substrate, the initial step
may be expected to be a rapid nucleation step involving the primary
specificity of the proteinase. Thus, in the case of pepsin, this
step may be considered to involve the Phe-Phe segment of the substrate, and to be followed by a cooperative process in which the
remaining segments of the oligopeptide are drawn into the site. The
rate of these subsequent interactions may be expected to depend on
the flexibility of the protein framework of which the active site
is a part.

The mechanism of pepsin action on synthetic substrates has been
the subject of extensive discussion, but many aspects of the problem
remain unresolved. From studies of the pH dependence of catalysis
(3) and chemical modification of presumed active-site groups (13), it
appears certain that at least two carboxyl groups of pepsin are
directly involved in the mechanism of the bond-breaking step. They
have been identified as belonging to Asp-32 and Asp-215 in the linear
sequence deduced for porcine pepsin (40), and it appears likely that
in the catalytically-active form of the enzyme one is protonated and
the other is in the form of the carboxylate anion. Studies on the
intra-molecular hydrolysis of phthalmic acid (41) and more recently,
the hydrolysis of dialkylmaleamic acids (42), have provided at attractive models for pepsin action. As in the general mechanisms
for the cleavage of amides and esters, a tetrahedral intermediate is
postulated, and it appears likely that this intermediate is formed
by nucleophilic attack at the carbonylcarbon of the sensitive bond.
The transformation of this intermediate appears to be subject to acid
catalysis, a conclusion supported by the similarity of the rates of
hydrolysis of the ester bond in Z-Phe(4NO$_2$)-Pla-OMe and of the
corresponding peptide (6). The transition state in the bond-breaking
step may be expected to resemble a tetrahedral intermediate having
the structure of an α-amino alcohol. Recent studies have shown
that suitable aldehydes are effective inhibitors of several proteinases, and it has been proposed that their hydrates are transition-
state analogs (43). In the case of pepsin it has been suggested (44)
that pepstatin acts by virtue of this principle, but the recent report that a synthetic deoxypepstatin retains the inhibitory power of
the natural inhibitor (45) casts doubt on the validity of this
suggestion.

Perhaps the most vexing questions about the mechanism of pepsin action on synthetic substrates relate to the order of the release products, and whether a covalently bound acyl-enzyme or amino-enzyme is an intermediate in the catalytic process. Since the rate-limiting step is associated with the decomposition of the first detectable enzyme-substrate complex, such intermediates do not accumulate (as in the cleavage of p-nitrophenyl acetate by chymotrypsin) and must be detected by means of trapping reactions. The decomposition of the presumed tetrahedral intermediate in pepsin catalysis, with the prior departure of the amine product, would be expected to leave behind an acyl-enzyme, but efforts to trap it with ^{14}C-labeled methanol have been unsuccessful (46). Aside from the well-established pepsin-catalyzed [18]O exchange of Ac-Phe with H_2[18]O, the principal evidence for the intermediate formation of a covalently-bound acyl-enzyme has come from the demonstration that substrates such as Leu-Tyr-Leu are cleaved by pepsin to form Leu-Leu (47,48). This process is assumed to involve the intermediate formation of a Leu-E intermediate which reacts with the substrate to form Leu-Leu-Tyr-Leu, whose cleavage yields Leu-Leu. Also, evidence for the existence of an acyl-enzyme has been adduced from studies on the cleavage of sulfite esters by pepsin (49); although these substrates appear to be cleaved at the active site of the enzyme (50), further studies are needed to determine whether the same active site groups are necessary for the scission of the sulfite and peptide substrates.

It has long been known that pepsin catalyses transpeptidation reactions with the apparent transfer of the amine product to an acidic acceptor, and this finding has been taken to indicate the prior departure of the acidic product and the formation of a covalently-bound amino-enzyme (3,13). Until recently, such transpeptidation reactions were observed largely with acyl dipeptides (e.g., Ac-Phe-Tyr) as substrates, and the corresponding esters or amides did not yield measurable amounts of transpeptidation product with ^{3}H-labeled Ac-Phe as the acceptor (51). It has been reported, however, that Ac-Phe-Phe-γ-propylmorpholinium ester does participate in a transpeptidation reaction with Z-Phe(NO_2) as the acceptor (9).

If a covalently-bound amino-enzyme is indeed an intermediate in pepsin catalysis, it would be expected that a series of substrates of the type A-Phe-Trp or A-Phe($4NO_2$)-Tyr, where the nature of the A group is varied, would yield the same intermediate (E-Trp or E-Tyr). It has been shown, however, that the ratio of labeled Ac-Phe-Trp formed in the presence of ^{14}C-labeled Ac-Phe to the Tryptophan formed by hydrolysis of A-Phe-Trp is not constant, and the substrate that is hydrolyzed most rapidly (A = Z-Ala-His) gave no detectable transpeptidation under the conditions of the experiments (52). Similarly, with substrates of the type A-Phe($4NO_2$)-Tyr (A = Z, Z-Gly, Z-Gly-Gly), the importance of secondary interactions is clearly evident, and the increased is correlated with the decrease in the relative effectiveness of Z-Phe($4NO_2$), Z-Gly-Phe($4NO_2$), and Z-Gly-Gly-Phe($4NO_2$) as acceptors in the transpeptidation reaction with Ac-Phe-Tyr as the substrate (53).

At present, therefore, the question of the ordered release of products, with the formation of detectable covalently-bound acyl-enzyme or amino-enzyme compounds, appears to be unresolved. It has been suggested (13) that such compounds may not be intermediates in pepsin catalysis, and attention has been drawn to the possible contribution of the conformational state of the active site to the order of the release of products. In light of the previous discussion of the possible relation of conformational changes at the active site of pepsin to its catalytic efficiency, it appears likely that after the bond-breaking step has occurred, the sequence of departure of the two products of hydrolysis may be a reflection of the interaction of each of them with complementary groups in the active site. These interactions may be coupled, so that the nature of one product may influence the rate of departure of the other through the effect it has on the conformational state of the active site, and the two products may therefore leave either a sequence that suggests apparent acyl transfer or in one that is consistent with apparent amino transfer. The possibility should be considered, therefore, that no detectable covalent acyl-enzyme or amino-enzyme intermediate can accumulate in pepsin-catalyzed reactions, but that the acidic product or the amine product (depending on its relative affinity for the active site) can stick to the active site longer than its partner. If this should be the case, either apparent acyl transfer or amine transfer would be possible by a direct condensation of the preferentially retained product and an acceptor that can readily displace the product that leaves more easily. That such condensation reactions are catalyzed by pepsin in the case of oligopeptides was demonstrated many years ago (54), and is consistent with the neglible free energy decrease in the hydrolysis of interior peptide bonds (55). For transpeptidation reactions in which an apparent E-Tyr amino-enzyme has been postulated, the free energy change in the condensation of an acceptor such as Ac-Phe with tyrosine would be more unfavorable in free solution, but the possibility must be considered that the ammonium pK_a of the tyrosine retained at the active site may be lower than that of tyrosine in free solution, perhaps by virtue of the interaction of the carboxylate group of the amino acid with a complementary cationic group of the active site.

Acknowledgments

The research of our laboratory reported in this article was aided by grants from the National Institutes of Health (GM-18172 and AM-15682) and from the National Science Foundation (BMS 73-06877).

REFERENCES

1. Fruton, J. S., and Bergmann, M. (1939) J. Biol. Chem. 127,
    627-641
2. Baker, L. E. (1951) J. Biol. Chem. 193, 809-819
3. Clement, G. E. (1973) Progr. Bioorg. Chem. 2, 177-238
4. Tang, J. (1965) J. Biol. Chem. 240, 3810-3815
5. Zeffren, E., and Kaiser, E. T. (1967) J. Amer. Chem. Soc. 89,
    4204-4208
6. Inouye, K., and Fruton, J. S. (1967) Biochem. 6, 1765-1777
7. Hollands, T. R., Voynick, I. M., and Fruton, J. S. (1969)
    Biochem. 8, 575-585
8. Sachdev, G. P., and Fruton, J. S. (1969) Biochem. 8, 4231-4238
9. Tikhodeeva, A. G., Rumsh, L. D., and Antonov, V. K. (1975)
    Bioorg. Khimia 1, 993-994
10. Trout, G. E., and Fruton, J. S. (1969) Biochem. 8, 4183-4190
11. Voynick, I. M., and Fruton, J. S., unpublished experiments
12. Rumsh, L. D., Tikhodeeva, A. G., and Antonov, V. K. (1974)
    Biokhimiya 39, 899-902
13. Fruton, J. S. (1976) Adv. Enzymol. 44, 1-36
14. Sampath-Kumar, P. S., and Fruton, J. S. (1974) Proc. Natl. Acad.
    Sci. U.S.A. 71, 1070-1072
15. Schechter, I., and Berger, A. (1967) Biochem. Biophys. Res.
    Commun. 27, 157-162
16. Lowbridge, J., and Fruton, J. S. (1974) J. Biol. Chem. 249,
    6754-6761
17. Sachdev, G. P., and Fruton, J. S. (1970) Biochem. 9, 4465-4470
18. Sachdev, G. P., Johnston, M. A., and Fruton, J. S. (1972)
    Biochem. 11, 1080-1086
19. Sachdev, G. P., Brownstein, A. D., and Fruton, J. S. (1973)
    J. Biol. Chem. 248, 6292-6299
20. Sachdev, G. P., Brownstein, A. D., and Fruton, J. S. (1975)
    J. Biol. Chem. 250, 501-507
21. Humphreys, R. E., and Fruton, J. S. (1968) Proc. Natl. Acad.
    Sci. U.S.A. 59, 519-525
22. Raju, E. V., Humphreys, R. E., and Fruton, J. S. (1972) Biochem.
    11, 3533-3536
23. Wang, J. L., and Edelman, G. M. (1971) J. Biol. Chem. 246,
    1185-1191
24. Sachdev, G. P., and Fruton, J. S. (1975) Proc. Natl. Acad. Sci.
    U.S.A. 72, 3424-3427
25. Press, E. M., Porter, R. R., and Cebra, J. (1960) Biochem. J.
    74, 501-514
26. Woessner, J. F., and Shamberger, R. J. (1971) J. Biol. Chem.
    246, 1951-1960
27. Voynick, I. M., and Fruton, J. S. (1971) Proc. Natl. Acad. Sci.
    U.S.A. 68, 257-259
28. Raymond, M. N., Garnier, J., Bricas, E., Cilianu, S., Blasnic, M.
    Chaix, A., and Lefrancier, P. (1972) Biochimie 54, 145-154

29. Fruton, J. S. (1975) in Proteases and Biological Control (Reich, E., Rifkin, D. B., and Shaw, E., Eds) pp. 33-50, Cold Spring Harbor Laboratory, New York
30. Inouye, K., and Fruton, J. S. (1967) J. Amer. Chem. Soc. 89, 187-188
31. May, S. W., and Kaiser, E. T. (1969) J. Amer. Chem. Soc. 91, 6491-6495
32. May, S. W., and Kaiser, E. T. (1972) Biochem. 11, 592-600
33. Hunkapiller, M. W., and Richards, J. H. (1972) Biochem. 11, 2829-2839
34. Takahashi, M., Want, T. T., and Hofmann, T. (1974) Biochem. Biophys. Res. Commun. 57, 39-46
35. Bender, M., and Kezdy, F. (1965) Ann. Rev. Biochem. 34, 49-76
36. Thompson, R. C., and Blout, E. R. (1973) Biochem. 12, 57-65
37. Jencks, W. P. (1975) Adv. Enzymol. 43, 219-410
38. Burgen, A. S. V., Roberts, G. C. R., and Feeney, J. (1975) Nature 253, 753
39. Mattis, J. A., and Fruton, J. S. (1976) Biochem. 15, 2191-2194
40. Sepulveda, P., Marciniszyn, J., Liu, D., and Tang, J. (1975) J. Biol. Chem. 250, 5082-5088
41. Bender, M. L., Chow, Y., and Chloupek, F. (1958) J. Amer. Chem. Soc. 80, 5380-5387
42. Kirby, A. J., McDonald, R. S., and Smith, C. R. (1974) J. Chem. Soc., Perkin Trans. 2 1974, 1495-1504
43. Wolfenden, R. (1976) Ann. Rev. Biophys. Bioeng. 5, 271-306
44. Marciniszyn, J. P., Hartsuck, J. A., and Tang, J. (1975) Fed. Proc. 34, 484
45. Rich, D. H., Sun, E., and Singh, J. (1976) Abstr. Amer. Chem. Soc. Divn. Biol. Chem., 172nd Meeting, San Francisco, Cal., No. 42
46. Cornish-Bowden, A. J., Greenwell, P., and Knowles, J. R. (1969) Biochem. J. 113, 369-375
47. Takahashi, M., and Hofmann, T. (1975) Biochem. J. 147, 549-563
48. Newmark, A. K., and Knowles, J. R. (1975) J. Amer. Chem. Soc. 97, 3557-3559
49. Nakagawa, Y., King Sun, L. H., and Kaiser, E. T. (1976) J. Amer. Chem. Soc. 98, 1616-1617
50. Zeffren, E., and Kaiser, E. T. (1968) Arch. Biochem. Biophys. 126, 965-967
51. Silver, M. S., and Stoddard, M. (1972) Biochem. 11, 191-200
52. Silver, M. S., Stoddard, M., and Kelleher, M. H. (1976) J. Amer. Chem. Soc. 98, (in press)
53. Richman, P. G., and Fruton, J. S. (1976) Proc. Natl. Acad. Sci. U.S.A. 73, (in press)
54. Determann, H., Heuer, J., and Jaworek, D. (1965) Ann. Chem. 690, 189-196
55. Kozlov, L. V., Ginodman, L. M., Orekhovich, V. N., and Valueva, T. A. (1966) Biokhymiya 31, 315-321

FOOTNOTE

*Abbreviations: Z, benzyloxycarbonyl; Ac, acetyl; Phe(4NO$_2$),
p-nitro-L-Phenylalanyl; Mns, mansyl, 6-(N-methylanilino)-2-
napthalene-sulfonyl; OP4P, 3-(4-pyridinium)propyl-1-oxy; Dns, dansyl,
5-dimethyl-aminonapthalene-1-sulfonyl; Pla, β-phenyl-L-lactyl; Tfa,
trifluoroacetyl; Nva, L-norvalyl; Nle, L-norleucyl; OMe, methoxy.
The abbreviated designation of amino acid residues denotes the L-
form.

SUBSITE SPECIFICITY OF PORCINE PEPSIN

James C. Powers, A. Dale Harley, and Dirck V. Myers

School of Chemistry
Georgia Institute of Technology
Atlanta, Georgia 30332

Corporate Research and Development Department
The Coca Cola Company
Atlanta, Georgia 30301

The specificity of porcine pepsin toward small synthetic substrates has been extensively investigated (1-4), but only a few
papers have dealt with the specificity of this enzyme toward polypeptide or protein substrates. Tang analyzed cleavage sites of 4
proteins by pepsin and concluded that the enzyme possessed a hydrophobic binding site (5). In addition, bonds split by pepsin in
seven peptides or proteins of established sequence have been summarized by Hill (6). In neither study was the sample size large
enough to yield any information other than the primary specificity
of pepsin. Antonov and his coworkers (7) recently reported a more
extensive analysis of pepsin specificity toward protein substrates
and concluded that the enzyme possessed five subsites.

To analyze the nature of the extended substrate binding site
of pepsin, we have collected and analyzed the sites of peptic cleavage
in a large number of proteins and polypeptides. The majority were
obtained by analyzing all papers published during 1967-1972 in
Biochemistry, J. of Biol. Chem. and Biochem. J. for proteins or polypeptides which had been enzymatically cleaved with porcine pepsin
during sequence determinations. The amino acid sequence of the
actual fragment treated with pepsin was entered into a computer along
with the site(s) of cleavage by the enzyme. In general, the proteins
were denatured with all disulfide bonds broken. Cyanogen bromide
peptides and fragments obtained by prior enzymatic treatment (e.g.
trypsin) were also included in the sample. Terminal amino acid

141

residues and chemically modified residues such as cysteic acid were
coded so they could be distinguished from an ordinary amino acid
residue.

The final population used in this study included 177 proteins
or peptides consisting of 7087 amino acid (AA) residues. The smallest
peptide was 5 residues in length and the largest protein contained
500 residues. The average length of the peptides was 40.1 residues,
with a median length of 20 AA residues. The original sample contained several proteins (cytochromes, immunoglobulins, keratin and
collagen) which were represented more than once due to the sequencing
of related proteins from different sources. This high degree of
homology in our original sample resulted in a biased distribution
of residues at secondary binding sites and lowered cleavage probabilities at the primary specificity site in a preliminary analysis.
Subsequently, only a representative number of each homologous group
of proteins was included in our sample and the data reported herein
were obtained from a sample which was as non-homologous as possible.

The sample contained 6910 peptide bonds, of which 1020 were
cleaved by pepsin; the cleavage probability thus was 1020/6910 =
0.148. This cleavage probability is in reality a cleavage frequency
observed within a particular sample of proteins and polypeptides.
Since we have a large representative sample, we have elected to term
our observed results as cleavage probabilities rather than cleavage
frequencies in order to indicate their potential applicability to
new systems. Cleavage probabilities have been obtained with a number
of other proteolytic enzymes:

Enzyme	Cleavage Probabilities
Chymotrypsin	0.14
Pepsin	0.15
Thermolysin	0.18
Elastase	0.24
Papain	0.26

Pepsin is obviously one of the more selective proteolytic enzymes in
terms of the overall number of cleavages catalyzed.

In our analysis, we have treated all bonds solely on the basis
of being split or not split, making no attempt to normalize such
factors as rates of reaction or percentage yields. With a large
sample, however, cleavage probabilities should be directly related
to rates of peptide bond cleavage. Those bonds which have high rates
of hydrolysis or which give high cleavage yields are most likely to
be noted as cleaved by individual investigators, no matter what
conditions were used. Conversely, those bonds with low hydrolysis

rates would more often be observed as not cleaved and would have low
cleavage probabilities.

The absolute cleavage probabilities, but not the relative probabilities, can be affected by the reaction conditions and by sampling
problems. A variety of conditions (pH, temperature, time, etc.) are
used in porcine pepsin digestions. If longer reaction times were
used by all investigators, all cleavage probabilities would be higher
and vice versa; but because of our large sample size, such differences
between separate investigations should cancel and our cleavage
probabilities are representative of those which should be obtained
under the conditions used by most sequence investigators. One sampling problem could affect the absolute probabilities: occasionally
a sequence investigator will report only the splits required for the
sequence determination. Since there is no reason to expect any bias
in which splits are reported and which are not reported, this sampling problem would affect only absolute probabilities. Thus the
cleavage probabilities which we report in this paper may be slightly
lower than actual cleavage frequencies; however, relationships and
rankings between individual amino acid residues will not be altered
either by this sampling problem or by the variety of conditions used
in pepsin digestions.

Discussion of the specificity of a proteolytic enzyme can no
longer be limited to the nature of the bond cleaved, but must also
include the effect of the various subsites on either side of the
cleaved bond. In discussing these subsites, we will use the nomen-
clature of Schechter and Berger (8); this is illustrated in Figure 1.

The cleavage probability of any particular peptide bond (AA_x-
AA_y) between two AA residues x and y is equal to number of AA_x-AA_y
bonds cleaved divided by the total number of AA_x-AA_y bonds in the
sample. For example, 4 Phe-His bonds out of a total of 5 in the
sample were cleaved by pepsin. This yields a cleavage probability
of 0.80 for Phe-His bonds. All bonds with a cleavage probability
of > 0.6 are listed in Table I. Terminal residues are not included
in the cleavage probabilities, but are discussed separately. Examination of the data shows that the majority of the peptide bonds with
high cleavage probabilities have aromatic or hydrophobic amino acids
as both the P_1 and P_1' residues. There are, however, a number of
examples where one of the AA residues is positively (His or Lys) or
negatively charged (Cys-Cm,Asp or Glu), although at the pH at which
most pepsin digestions occur, Glu and Asp are likely to be only
partially ionized.

The preference of pepsin for peptide bonds involving two aromatic
or hydrophobic AA residues are previously observed by Tang (5) and
with synthetic substrates by various investigators (see Chapter 8 in
this volume). This is in striking contrast to chymotrypsin and

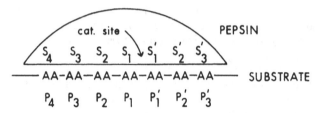

Figure 1. A representation of the interaction of polypeptide substrate with the extended substrate binding site of pepsin. The individual amino acid (AA) residues are designated P_1, P_2, P'_1, etc. and the corresponding subsites of the enzyme are S_1, S_2, S'_1, etc. The bond cleaved by the enzyme is the peptide bond between the P_1 and P_2 residues.

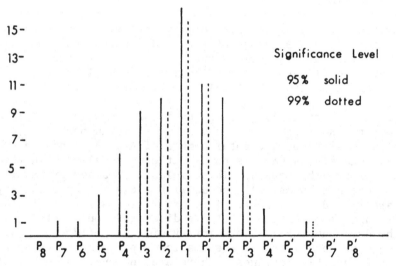

Figure 2. The number of individual amino acid residues whose cleavage probability by pepsin is significantly different (greater or smaller) than the mean cleavage probability at each of the subsites P_8-P'_8. The solid line indicates the number which differs at the 95% confidence level and the dotted line, those which differ at the 99% confidence level using a chi-squared test.

Table I

Peptide Bonds with High Cleavage Probabilities

Cleavage Probability > 0.8			
Phe-Tyr	Phe-His	Phe-Ala	Met-Met
Met-His	Trp-Glu	Trp-Ile	Cys(Cm)-Tyr
Tyr-Met			
Cleavage Probability > 0.7			
Phe-Val	Asp-Phe	Asp-Tyr	
Cleavage Probability > 0.6			
Phe-Phe	Met-Leu	Met-Ser	Met-Lys
Leu-Phe	Leu-Leu	Trp-Phe	Trp-Tyr
Trp-Ala	Trp-Lys	Cys(Cm)-Val	

thermolysin where only one subsite, S_1 and S_1' respectively, is the primary determinant of cleavage.

The cleavage probabilities for all amino acid residues at all subsites were also determined in the current analysis. The definition of the individual subsite probabilities is illustrated with the AA residue alanine in Table II. The subsite cleavage probabilities varied from 0.51 (Phe at P_1) to 0 (aminoethylcysteine at P_1). The subsite probabilities for all major amino acid residues at P_1 and P_1' are listed in Table III. The S_1 subsite of pepsin is obviously the primary determinant of specificity, since the probabilities are much higher at this subsite than at S_1' (or at any other subsite). The AA residues with high cleavage probabilities at P_1 are Phe, Met, Leu and Trp; those with high probability at P_1' are Tyr, Phe and Ile.

We were interested in determining which subsite probabilities were significantly greater or lower than the mean (0.148) or the cleavage probability for any peptide bond. To accomplish this, we calculated chi-squared (χ^2) values:

$$\chi 2 = \frac{(f_o - f_e)^2}{f_e} + \frac{(f_o - f_e)}{f_e'}$$

f_o = number of a particular amino acid residue observed to occur at a given subsite in the population of cleaved bonds.

f_e = number of that amino acid residue expected to occur at a given subsite in the population of cleaved bonds if random cleavage (0.148 probability) was observed.

f_o', f_e' = corresponding number of a particular amino acid residue expected to occur in the non-cleaved population.

The number of amino acid residues which had cleaved probabilities significantly different than the mean by the chi-squared test at both the 95% ($\chi^2 > 3.84$) and 99% ($\chi^2 > 6.64$) confidence level are plotted in Figure 2 for 16 subsites (P_8-P_8'). Examination of the figure shows once again that S_1 is the primary specificity site of pepsin since more amino acid residues differ from the mean at P_1 than at any other subsite. The length of the extended substrate binding site can also be determined from Figure 2. At subsites outside the extended substrate binding site of the enzyme, where there is no interaction between the enzyme and the substrate, all amino acid residues would be expected to have cleavage probabilities which do not differ significantly from the mean. This is in fact the case at subsites such as P_8, P_5', P_7' and P_8'. At some other subsites such as P_7, P_6, P_6' and P_4',

Table II

Subsite Cleavage Probabilities for Alanine[a]

P_1 cleavage probability $= \dfrac{\text{-Ala*AA- bonds cleaved}}{\text{-Ala-AA- bonds total}} = \dfrac{82}{502} = 0.16$

P_3 cleavage probability $= \dfrac{\text{-Ala-AA-AA*AA- bonds cleaved}}{\text{-Ala-AA-AA-AA- bonds total}} = \dfrac{93}{468} = 0.20$

P_2 cleavage probability $= \dfrac{\text{-AA*AA-Ala- bonds cleaved}}{\text{-AA-AA-Ala- bonds total}} = \dfrac{89}{482} = 0.19$

---

[a]The asterisk indicates the bond cleaved. AA represents any amino acid residues. Terminal amino acid residues were considered separately and are not included in the subsite probabilities.

Table III

Subsite Cleavage Probabilities at P_1 and P_1'[a]

AA Residue	Total Occurrence[b]	P_1	P_1'
Phe	231	.51	.29
Met	68	.43	.18
Leu	461	.41	.20
Trp	63	.40	.21
Asp	310	.26	.16
Glu	338	.24	.13
Tyr	235	.24	.34
Ala	502	.16	.21
	mean 0.148 $\longrightarrow$		
Gln	278	.14	.08
Asn	237	.12	.08
Cys(Cm)	205	.11	.11
Thr	487	.11	.12
Ser	659	.09	.08
Val	494	.08	.23
Gly	566	.07	.08
His	140	.06	.11
Arg	239	.06	.11
Ile	304	.05	.26
Lys	309	.03	.15
Pro	399	.01	.09

[a]The amino acid residues are listed in descending order of P_1 cleavage probability.

[b]This refers to the P_1 subsite. It would be slightly different at P_1'.

only a few AA residues have significantly different cleavage probabilities. We believe this is "notice" in the data simply because all
homologies have not been completely eliminated from the sample. If
one accepts this limitation of the plot and considers only those subsites which have significant numbers of amino acid residues with
probabilities different from the mean at the 99% confidence level,
then the conclusion is reached that pepsin has a 7-subsite binding
site. The extended binding site must involve subsites S_4-S_3' and
interaction with AA residues P_4-P_3' of a substrate (Fig.1). It should
be noted that any interaction of the enzyme with only the backbone of
an amino acid residue of substrate would not be detected. The
cleavage probabilities would differ from the mean only if the enzyme
were in some way "seeing" or recognizing the side chain of individual
amino acids at a particular subsite. Thus, although it is possible
that the extended substrate binding region is greater than 7 subsites,
we consider it unlikely. Antonov et al., on the basis of a similar
statistical analysis on a smaller sample, have concluded that pepsin
has only a 5-subsite (P_3-P_2') binding site (7). A 7-subsite extended
substrate binding site is certainly reasonable for an enzyme of pepsin's size, and we are eager to see whether crystallographic studies
confirm our expectations.

All the amino acid residues which differ significantly from the
mean at subsites P_4-P_3' are shown in Figure 3. Examination of the data
shows that at certain subsites, a group of related AA residues is
favored or disliked by pepsin. For example at P_3', a long alkyl side
chain (Leu or Ile) is preferred. At P_1', alkyl side chains of any
length and/or aromatic groups (Phe or Tyr) interact favorably with
the enzyme. Rather than discussing all of the subsite interactions,
we have chosen to restrict our attention to a few of the more significant. Note that favorable cleavage probabilities could be due to
better binding (lower K_m), better catalysis (higher k_{cat}) or both.
Our treatment does not distinguish these possibilities.

Proline is one amino acid which would be expected to profoundly
influence the reactivity of a peptide substrate. Proline restricts
the possible conformations of a peptide chain and in addition is unable to act as a hydrogen bond donor. Both of these factors could
affect peptide bond cleavage by hindering substrate binding or by
preventing proper catalysis. Alternately, favorable interaction
could take place between the enzyme and a prolyl residue, due either
to a favorable hydrophobic interaction with the prolyl side chain or
because proline restricts the peptide conformation to one which is
favorable. That these effects are important is accentuated by the
fact that at every subsite the cleavage probability of a substrate
with a proline residue is significantly different from the mean
(0.148). This is true for no other amino acid residue. Proline is
favorable at P_4 and P_3 and unfavorable at all other subsites (P_2-P_3').

P_4	P_3	P_2	P_1	P_1'	P_2'	P_3'
Asn (.22)	Ile (.20)	Glu (.27)	Phe (.51)	Tyr (.34)	Val (.24)	Leu (.22)
Pro (.21)	Ala (.20)	Asn (.22)	Met (.43)	Phe (.29)	Arg (.23)	Ile (.19)
Tyr (.21)	Thr (.20)	Ser (.22)	Leu (.41)	Ile (.26)	Ala (.19)	
Ile (.20)	Gly (.20)	Val (.21)	Trp (.40)	Val (.23)	Glu (.19)	
Asp (.19)	Pro (.19)	Ala (.19)	Asp (.26)	Ala (.21)	Ile (.19)	
			Glu (.24)	Leu (.20)	Thr (.18)	
			Tyr (.24)			

P_4	P_3	P_2	P_1	P_1'	P_2'	P_3'
Cys(Cm) (.09)	Phe (.10)	Gly (.11)	Cys(Cm) (.11)	Pro (.09)	Leu (.10)	Ser (.12)
	His (.07)	Leu (.09)	Thr (.11)	Gly (.08)	Gly (.10)	Gly (.11)
	Lys (.07)	Phe (.06)	Ser (.09)	Ser (.08)	Phe (.09)	Pro (.07)
	Arg (.02)	Pro (.05)	Val (.08)	Gln (.08)	Pro (.04)	
		Tyr (.04)	Gly (.07)	Asn (.08)		
			His (.06)			
			Arg (.06)			
			Ile (.05)			
			Lys (.03)			
			Cys(Ae) (0)			
			Pro (.01)			

Figure 3. Subsite preferences and dislikes for pepsin. Those amino acid residues listed on the top have cleavage probabilities greater than the mean at the 95% confidence level. Likewise, those listed on the bottom have cleavage probabilities less than the mean.

Table IV

Effect of Basic Residues at P_3 in a Polypeptide Substrate

	number of bonds	expected cleavage[a]	actual cleavage	cleavage probability
-His-AA-AA*AA-[b]	131	19.4	9	.0687
-Lys-AA-AA*AA-[b]	291	43.1	19	.0653
-Arg-AA-AA*AA-[b]	232	34.1	4	.0172

[a]Based on the mean cleavage probability of 0.148.

[b]The asterisk indicates the site of cleavage.

Another significant subsite effect is observed with basic residues at P$_3$. Table IV shows the negative effect of His, Lys, and Arg residues at P$_3$ in polypeptide substrates on the cleavage probability. Arginine in particular hinders peptide bond cleavage. This was even more striking when only very susceptible bonds, those involving Phe, Met, Leu or Trp at P$_1$, were examined (Table V). In this group, the cleavage probability drops from 0.436 to 0 when an Arg occurs at P$_3$. Similar results were observed with His and Lys, although the probabilities were never zero.

Two explanations for the effect of basic residues at P$_3$ can be imagined (Fig.4). In the first, the S$_3$ subsite of pepsin might contain some structural feature such as a positively charged group or hydrophobic 'niche' which prevents substrate binding. We consider this possibility unlikely; the S$_3$ subsite can accommodate negatively charged groups (Glu and Asp) and so this region of the enzyme cannot specifically require a non-charged residue. Even if this were not the case, there are several examples where basic AA side chains can occupy hydrophobic subsites of enzymes (9). A more compelling argument is that, for the majority of subsites in pepsin and other proteases, most of the amino acid residues are either favorable or neutral. Due to the variety of side chain functional groups, which most subsites will accept, we find it hard to imagine how a subsite, especially one removed from the catalytic site, could be structured to so completely exclude one group of amino acids.

An alternate and more likely explanation is that a basic residue at P$_3$ in a substrate induces a conformational change in pepsin which inactivates the catalytic residues (Fig.4). It is also possible for the substrate to be displaced in some way with respect to the catalytic site (non-productive binding). Since there are several carboxyl groups in the catalytic site of pepsin which are thought to be catalytically involved, it is possible that a basic residue at P$_3$ in a substrate forms an ionic bond with one of these carboxyl groups, inactivating the enzyme. In this regard, one should note that basic residues at P$_1$ also have low cleavage probabilities, possibly for the same reason. One final observation should be made. For crystallographers interested in studying the nature of the extended substrate binding site of pepsin and possibly homologous acid proteases, a peptide with a sequence such as -Arg-AA-Phe-AA- would probably form a stable complex with pepsin. A similar non-productive complex of Gly-Tyr with carboxypeptidase A has been extremely informative (10).

It is well known that pepsin is an endopeptidase, so we expected terminal residues at P$_1$ and P$_1'$ to lower the cleavage probability substantially. This is indeed the case, as can be seen in Table VI. It is clear also that a NH$_2$-terminal residue at P$_3$ has almost no effect, while one at P$_2'$ hinders cleavage. COOH-terminal residues at P$_2'$ also hinder hydrolysis. The effect of terminal residues at other subsites was not examined.

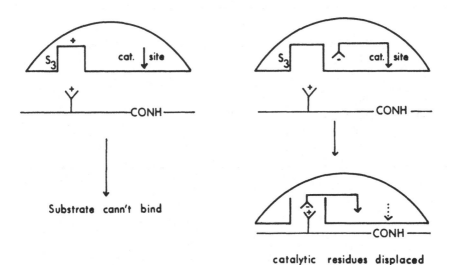

Figure 4. Possible explanations for the negative influence of a basic residue at P_3 in a polypeptide substrate on the cleavage probability. On the left, the S_3 subsite of pepsin may contain a positively charged residue or some other structural feature which prevents binding of the substrate. On the right, binding of a substrate with a P_3 basic residue causes a conformational change resulting in inactivation of the catalytic residues.

Table V

Effect of an Arginine Residue at P_3 in Polypeptide
Substrates Which are Highly Susceptible to Cleavage

	number of bonds	cleavage probability
Phe -AA-AA-Met*AA- Leu Trp	823	0.436
Phe -Arg-AA-Met*AA- Leu Trp	29	0

Table VI

Effect of Terminal Residues on Pepsin Cleavage Probabilities[a]

Terminal Positions	Probabilities	Terminal positions	Probability
-HPA*AA-	0.44	AAZ-HPA*AA	0.11
HPA*AA-	0.18	AAZ-AA-HPA*AA-	0.47
-HPA*AAZ	0.11	-HPA*AA-AAZ	0.29

[a]AA = any amino acid residue. HPA = an amino acid residue
with a high cleavage probability (Phe, Met, Leu and Trp).
AAZ = any terminal amino acid.

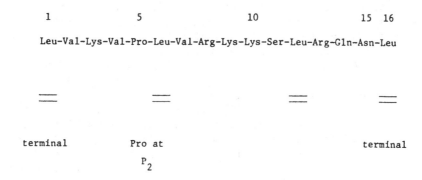

terminal   Pro at   terminal
                    P_2

P_4 P_3 P_2 P_1 P_1' P_2' P_3'

-Lys-Lys-Ser-Leu-Arg-Gln-Asn-

n   (-)   (+)   (+)   n   n   n

Figure 5. The sequence of a pepsin-inhibiting peptide formed
during the activation of porcine pepsinogen. The lower sequence
illustrates the proposed binding mode with Leu-12 as the P_1
residue. Below each residue is listed whether that amino acid
residue has positive (+), negative (-), or neutral (n) cleavage
probabilities.

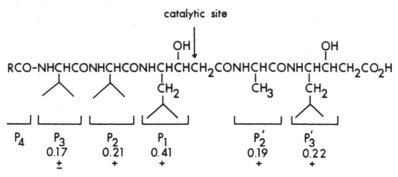

Figure 6. Structure of pepstatin. The subsites are designated
with the assumption that the side chain of the central 4-amino-3-
hydroxy-6 methylheptanoic acid residue binds at the S_1 subsite of
pepsin. Cleavage probabilities are listed under each amino acid
residue. Favorable (+) and neutral (±) probabilities are indicated.

As a test of our cleavage probability data, we sought to apply it to new systems. One interesting example is a pepsin-inhibiting peptide formed during the activation of both porcine and bovine pepsinogen. It corresponds to the NH_2-terminal 16 amino acids of porcine pepsinogen (11,12) and the terminal 17 amino acids of bovine pepsinogen (13). We examined the sequence of this peptide (Fig.5) to see if we could explain its inhibitory properties. The two peptides contain 4 leucyl residues (P_1 cleavage probability = 0.51). The amino acid with the next highest P_1 cleavage probability is Gln with 0.14. Two of the leucyl residues are terminal and will not interact well with pepsin (see Table VI). The third (Leu-6) has a prolyl residue it P_2, an unfavorable interaction. The fourth leucine residue (Leu-12) is part of a sequence (Fig.5) where each of the 7 subsites has an amino acid residue which is either favorable or neutral except one. There is a Lys at P_3, and we propose that this region of the peptide inhibits pepsin because the peptide either binds nonproductively or induces a conformational change in the enzyme.

Pepstatin is a strong competitive inhibitor of pepsin which is produced by various strains of actinomycetes (14). The structure (Fig.6) contains an unusual amino acid; 4-amino-3-hydroxy-6-methyl-heptanoic acid (statine). Tang has proposed that the statine residue of pepstatin resembles the transition state for pepsin hydrolysis of peptide bonds (15, see also Chapter 12 in this volume). If this is so and the central statine residue binds at the catalytic site with its side chain in the S_1 subsite, then we note (Fig.6) that the interactions at all the other subsites are favorable or neutral. The P´ subsites are listed with the assumption that central residue bridges two subsites (S_1 and S_1').

In conclusion, we should note that the data reported in this paper should be valuable for the design of new substrates and inhibitors for pepsin.

ACKNOWLEDGEMENT

This research was supported by U.S. Public Health Service Grant GM 18292.

SUMMARY

Although the specificity of pepsin has been extensively investigated using small synthetic substrates, the action of pepsin on polypeptide substrates is less throughly studied. We examined 3 major biochemical journals during 1967-1972 for all papers published regarding proteins or peptides which were cleaved by pepsin in the course of sequence studies. The cleavages were than analyzed to determine the subsite specificity of pepsin. The population consisted

of 6910 peptide bonds of which 1020 (14.8%) were cleaved by the
enzyme. Bonds with high probability of cleavage (>0.70) were
Phe-Val, Asp-Phe, Asp-Tyr, Phe-Tyr, Phe-His, Met-Met, Met-His,
Trp-Glu, Phe-Ala, Cys(cm)-Tyr, Tyr-Met, and Trp-Ile. Analysis of
the specificity by subsites showed that P_1 was the primary determinant of pepsin specificity with hydrophobic amino acids Phe (probability = 0.51), Met (0.43), Leu (0.41), and Trp (0.40) being preferred. The extended substrate binding site in pepsin is composed
of seven subsites P_4-P_3'. At subsites other than P_1 most amino acid
residues have no affect on the cleavage probability, but at each
subsite there are some amino acid residues which exert a favorable
or unfavorable influence on cleavage. For example, proline is
favorable (higher probability of being adjacent to a cleaved bond)
at P_4 and P_3 and unfavorable at the rest of the subsites. Arg is a
very unfavorable residue at P_3 with a 0.02 probability of cleavage
of an -Arg-AA-AA*AA-bond (cleavage at the asterick, AA = any amino
acid). In fact, among the 29 bonds in the sample with an Arg at P_3
and a highly susceptible residue (Phe, Met, Leu, Trp) at P_1, none
were cleaved. The effect of terminal residues was also analyzed.
A terminal residue at P_1, P_1', P_2, or P_2' considerably reduces the
cleavage probability; however, a terminal residue at P_3 has little
affect. Cleavage probabilities were used to predict the mode of
interaction of pepstatin and pepsin inhibiting activation peptide
with pepsin.

REFERENCES

1. Fruton, J. S. (1970) Adv. Enzymol. 33, 401-443
2. Fruton, J. S. (1971) The Enzymes (3rd edition) 3, 119-164
3. Fruton, J. S. (1976) Adv. Enzymol. 44, 1-36
4. Clement, G. E. (1973) Prog. in Bioorg. Chem. 2, 177-238
5. Tang, J. (1963) Nature 199, 1094-1095
6. Hill, R. L. (1965) Adv. Protein Chem. 20, 37-107
7. Zinchenko, A. A., Rumsh, L. D., and Antonov, V. K. (1976)
    Bioorg. Khim. 2, 803-810
8. Schechter, I., and Berger, A. (1967) Biochem. Biophys. Res.
    Commun. 27, 157-162
9. Poulous, T., Alden, R. A., Freer, S. T., Birktoft, J. J., and
    Kraut, J. (1976) J. Biol. Chem. 251, 1097-1103
10. Hartsuck, J. A., and Lipscomb, W. N. (1971) The Enzymes (3rd
    edition) 3, 1-56
11. Anderson, W., and Harthill, J. E. (1973) Nature 243, 417-419
12. Ong, E. B., and Perlmann, G. E. (1968) J. Biol. Chem. 243,
    6104-6109
13. Harboe, M., Andersen, P. M., Foltmann, B., Kay, J., and Kassell,
    B. (1974) J. Biol. Chem. 249, 4487-4494
14. Umezawa, H. (1972) in Enzyme Inhibitors of Microbial Origin,
    pp. 15-52, Baltimore, Md., University Park Press.
15. Tang, J. (1976) Trends in Biochem. Sci. 1, 205-208

ANHYDRIDE INTERMEDIATES IN CATALYSIS BY PEPSIN:

IS PEPSIN AN ENZYME WITH TWO ACTIVE SITES?

E. T. Kaiser and Y. Nakagawa

Departments of Chemistry and Biochemistry
University of Chicago
Chicago, Illinois 60637

Porcine pepsin is known to have two types of activities in acidic solution: the catalysis of peptide bond hydrolysis and the catalysis of transpeptidation (1). The presence of amino-enzyme intermediates in the catalytic action of pepsin has been proposed by Knowles (2-4) and Antonov et al. (5). On the other hand, Silver (6) and Hofmann (7) have proposed the existence of acyl-pepsin intermediates. Some mechanisms for pepsin action such as that suggested by Zeffren and Kaiser (8) involve the postulation of both acyl- and amino-enzyme intermediates. Recently, Fruton (9,10) has suggested that kinetically significant conformational changes of the pepsin active site in the presence of substrates or products instead of the formation of acyl-enzyme or amino-enzyme intermediates may account for the various data obtained for peptide hydrolysis and transpeptidation.

Fahrney and Reid (11) discovered that pepsin catalyzes the hydrolysis of sulfite esters. Extensive kinetic studies on these reactions have been performed in our laboratory (12-15). The following observations suggest that the active sites involved in the hydrolysis of sulfite esters and in the hydrolysis of peptides may be identical, or may at least overlap:

a) N-Acetyl-L-phenylalanyl-L-3,5-dibromotyrosine or its constituent amino acid derivatives acted as competitive inhibitors toward the sulfite esterase activity of pepsin; the K_i values measured corresponded closely with K_m or K_i values obtained from peptide hydrolysis data (12).

b) Pepsin modified by reaction with the active site-directed esterifying reagent N-diazoacetyl-D, L-norleucine methyl ester was not

only inactivated as a peptidase, but also as an esterase in catalyzing the hydrolysis of diphenyl sulfite and methyl phenyl sulfite at
pH 2 (11).

c) The pH profile of k_{cat}/K_m for the hydrolysis of bis-p-nitrophenyl catalyzed by pepsin showed a bell-shape, and ionizing groups
on the enzyme of pK = 0.82 and pK = 5.17 appeared to be involved in
this reaction. These parameters were close to those for the pepsin-
catalyzed hydrolysis of a neutral dipeptide substrate N-acetyl-L-
phenylalanyl-L-phenylalanine amide (14,16).

The presence of anhydride intermediates during the course of
the hydrolysis of sulfite esters catalyzed by pepsin was proposed
by May and Kaiser (14). Studies of the catalysis of sulfite ester
hydrolysis by model carboxylate species indicated that the presence
of anhydride intermediates could be detected in such reactions by
the use of nucleophilic trapping reagents (17). Based on the results
of the model studies, we were encouraged to attempt to trap the hypothetical anhydride intermediates formed in the pepsin-catalyzed
hydrolysis of a sulfite ester using hydroxylamine as the trapping
agent, which could lead to the identification of the active sites
involved in this reaction.

EXPERIMENTAL

Materials

Hemoglobin substrate powder and twice-crystallized porcine pepsin (PM 2LB and PM 8HA) were purchased from Worthington. Hydroxylamine hydrochloride was recrystallized from methanol. Benzohydroxamic
acid was prepared by the method of Hauser and Renfrow (18), mp 127-
128° (uncorr. lit. mp 131°). Acetonitrile was purified according to
a method of O'Donnell et al. (19). Phenyl tetrahydrofurfuryl sulfite
was a gift from Dr. L. -H. King Sun.

α-Naphthylamine was recrystallized several times from ligroin
until white needles were obtained and kept in a sealed amber bottle.
2,3-Diaminopropionic acid and 2,4-diaminobutyric acid were obtained
from Calbiochem. Sephadex G-25 was a product of Pharmacia Fine
Chemicals. 2,4-Dinitrofluorobenzene was purchased from Eastman-
Kodak. Other chemicals were the best grade available and used without further purification.

Methods

Purification of pepsin. Commercial pepsin (50 to 100 mg) was
purified by passage through a column of Sephadex G-25 (2 x 75 cm)

which was equilibrated with 0.01 M 2-(N-morpholino)ethane sulfonic acid (MES), pH 5.3.

Hemoglobin assay. Pepsin activity was determined by the method of Chow and Kassell (20).

Analysis of hydroxamic acid. Hydroxamic acid was analyzed by the method of Bergmann and Segal (21).

Trapping of the intermediates. Hydroxylamine hydrochloride was dissolved to a concentration of 0.01 M in a solution containing freshly purified porcine pepsin (usually about 3 mg/ml, total volume of 15 to 20 ml) in 0.01 M MES buffer, pH 5.3 and the pH was readjusted to 5.3, if necessary. Phenyl tetrahydrofurfuryl sulfite (PTFS) was dissolved in CH_3CN and the required amount of the stock solution was added to the pepsin solution at room temperature. Control experiments were carried out in parallel in the presence of CH_3CN only. One incubation cycle was for 30 min, and additional substrate solution was added at 30 min intervals for multiple incubation experiments. At the end of the experiments, the enzyme solution was dialyzed at 4^0 against five changes of 3 l of 0.01 M sodium acetate, pH 4.0, for a total of 72 hr. Measurements of peptic activity by the hemoglobin assay method and the quantitative analysis of hydroxamic acid were performed as mentioned above.

Lossen rearrangement and identification of the amino acid involved in intermediate formation. The dinitrophenylation method and Lossen rearrangement procedure followed the description of Gross and Morrell (22). The solution of hydroxamate modified was adjusted to pH 8 by the use of sodium bicarbonate and incubated with an equal amount of 1% 2,4-dinitrofluorobenzene in ethanol for 30 min at room temperature. After the reaction, the excess of 2,4-dinitrofluorobenzene was extracted with ether several times, until no yellow color was observed. The protein was dissolved in 0.1 N NaOH and the resultant solution was immersed in boiling water for 10 min. After acidification, the product was lyophilized, then hydrolyzed in a sealed tube containing 6 N HCl for 24 hr at 110^0 (23). The amino acid analysis was performed on a 2 x 53-cm column of PA-15 in a Beckman amino acid analyzer Model 121, eluting with 0.35 N sodium citrate, pH 3.8, at 30^0. The peak of 2,3-diaminopropionic acid appeared at 231 min, and 2,4-diaminobutyric acid was eluted at 240 min.

Polyacrylamide gel electrophoresis. Standard 7.5% polyacrylamide gels were prepared and run in glycine-Tris buffer, pH 9.0 at a constant current of 25 mA at room temperature. Bromophenol Blue was used as a marker and Coomassie Brillian Blue was the staining reagent.

Circular dichroism (CD) spectra. Pepsin samples were dissolved in 0.01 M sodium acetate, pH 4.0, and CD spectra were measured on a

Cary 60 spectrophotometer with the Model 6001 CD accessory. Fused
quartz cells with 10 mm and 1.0 mm light pathlengths were used.

Chemical modifications of pepsin. Purified pepsin was subjected
to treatment with the following previously investigated modifying
agents: (a) Diazoacetyl-D, L-norleucine methyl ester (24); (b) α-
diazo-p-bromoacetophenone (25); (c) 1,3,-epoxy-3-(p-nitrophenoxy)-
propane (26); (d) a combination of reagents (b) and (c).

The numbering of the amino acids in porcine pepsin used in this
manuscript follows Tang's sequencing results (27,28).

RESULTS

Purification of pepsin. At the beginning of this study, commercial pepsin was purified by passage through a column of hydroxyapatite
with a stepwise increase in the sodium phosphate concentration at pH
5.6, and multiple pepsin peaks were separated as Rajagopalan et al.
reported (29). In our latest work, commercial pepsin was purified by
passage through a column of Sephadex G-25 (2 x 75 cm) which had been
equilibrated with 0.01 M MES buffer, pH 5.3 by a procedure similar to
the one described by Cornish-Bowden and Knowles (16). When this
method was used, only two peaks, the enzymatically active major peak
and an inactive minor peak, were separated and the specific activity
of the purified pepsin was similar to that of pepsin purified by the
hydroxyapatite method.

Trapping experiments. The number of moles of hydroxamic acid
found per mole of pepsin after incubation with various amounts of
PTFS in 0.01 M hydroxylamine at pH 5.3 are summarized in Table I.
At a molar ratio of PTFS/pepsin = 2.5 in a single incubation, 20%
of the peptic activity was lost with incorporation of 1 mole of
hydroxamate per mole of pepsin. With increases in the number of
incubations or increases in the concentration of PTFS, the enzymatic activity decreased and the number of hydroxamates incorporated
per mole of pepsin increased. The maximum number of hydroxamate
groups found was 4 moles per mole of pepsin, and approximately 50%
of the peptic activity was lost.

Identification of the site of hydroxamate incorporation. To
identify which amino acids were involved in the incorporation of
hydroxylamine and the concomitant loss of the enzyme activity, all
modified pepsins containing from 1.2 to 3.8 mole of hydroxamate
per mole of pepsin (Table I) were subjected to Lossen rearrangement,
followed by acid hydrolysis. All samples contained 2,3-diamino-
propionic acid, but no 2,4-diaminobutyric acid was found using the
amino acid analyzer. Thus, the PTFS-pepsin intermediates trapped
with hydroxylamine must have been formed at the β-carboxyl groups of
aspartate residues and glutamate residues were not involved.

Table I

Hydroxamate Incorporation and Remaining Catalytic Activity in the
Catalyzed Hydrolysis of Phenyl Tetrahydrofurfuryl Sulfite in the
Presence of Hydroxylamine[a,b]

Mole ration of PTFS to pepsin	Number of incubations[c]	Remaining pepsin activity	Moles of hydroxamate per mole of pepsin
2.5	1	81.9%	1.2
2.2	2	61.2	3.3
5.0	1	59.5	1.4
5.0	2	52.0	3.8
9.2	1	57.2	1.4
11.7	9	53.3	3.2

[a]Solutions containing porcine pepsin (Worthington PM 8HA and
PM 2LB) were purified by passing them through a column of Sephadex
G-25 which had been equilibrated with 0.01 M 2-(N-morpholino)ethane-
sulfonic acid (MES) buffer, pH 5.3. Phenyl tetrahydrofurfuryl sulfite
(PTFS) in CH_3CN was added to the pepsin solution containing 0.01 M
hydroxylamine (pH 5.3), and the resultant solution was equilibrated
for 30 min at room temperature. Control experiments for the measure-
ment of pepsin activity were carried out in parallel in the presence
of CH_3CN only. Experiments on the incubation of hydroxylamine with
pepsin in the absence of the sulfite ester revealed no significant
incorporation of hydroxylamine during time periods comparable to
those in the experiments summarized in this table.

[b]This table has been reproduced from reference 30 with the permission of the American Chemical Society.

[c]Additional substrate was added at 30 min intervals for multiple
incubation experiments. At the end of the experiments, the enzyme
solution was dialyzed against 3 l of 0.01 M sodium acetate buffer,
pH 4.0, for 72 hr with 5 changes. Hydroxamic acid was quantitatively
analyzed by the method of Bergmann and Segal, (21). Pepsin activity
was determined by the hemoglobin assay method described by Chow and
Kassell (20).

Since our experiments demonstrated that anhydride formation in the pepsin-catalyzed hydrolysis of PTFS involved the β-carboxyl groups of aspartate, we set out to identify the positions of these residues in the amino acid sequence of the enzyme. A solution of pepsin containing 3.3 mol of hydroxamate per mole of enzyme (a total of 100 mg of modified pepsin, see the second entry in Table I) was adjusted to pH 3.5 with 0.1 M acetic acid, and 5% of pepsin (by weight) was added to the solution. The mixture was incubated at 25° for 20 hr, then passed through a column of Sephadex G-25 (2 x 88 cm) with water as the eluent. The effluent was monitored at 220 nm and the fractions containing digested peptide fragments were collected. Peptide fractions were subjected further to the following sequence of isolation steps:

a) High-voltage electrophoresis at pH 6.5 (pyridine-acetic acid-water system), 3,000 V for 1 hr on Whatman 3 mm paper using Savant immersion type high-voltage electrophoresis equipment.

b) High-voltage electrophoresis at pH 3.6 (pyridine-acetic acid-water system), 3,000 V for 1 hr.

c) Descending paper chromatography in t-butanol-methyl ethyl ketone-water (2:2:1, v/v/v) on Whatman No. 1 paper.

After each separation step, chromatograms were scanned under short wavelength uv light and hydroxamate positive zones were located by the method of Bergmann and Segal (21). The hydroxamate positive portions were eluted from the final descending paper chromatogram with water and hydrolyzed in 6 N HCl for 20 hr for amino acid analysis.

The results of amino acid analysis of the hydroxamate-positive zones are shown in Table II (30). The first column giving the results for two fractions (P-II-A-1-a and P-II-B-1-a) indicates an amino acid composition of Asp_1, Thr_1, Ser_1, Glu_1, Gly_3, Ala_1, (Cys), Val_1, Ile_1 and Leu_1. Upon examining the amino acid sequence of pepsin (27,28), it appears that the composition obtained is consistent with postulating that the fractions correspond to the following fragment in the vicinity of Asp-215, the group modified by diazocarbonyl reagents (24,25,28):

```
      210                    215                      220
Gly-Gly-Cys-Gln-Ala-Ile-Val-Asp-Thr-Gly-Thr-Ser-Leu
      |___|
```

The fractions listed in the second column of Table II (P-II-A-1-b and P-II-B-1-b) are composed of Asp_1, Thr_1, Ser_1, Gly_1, Val_1, Ile_1 and Phe_1. This fragment appears to correspond to the following sequence which includes the active site residue Asp-32, which was modified by 1,2-epoxy-3-(p-nitrophenyl)propane (26):

```
      29          32
Val-Ile-Phe-Asp-Thr-Gly-Ser
```

Table II

Amino Acid Composition of Isolated Hydroxamate-Positive Peptides[a]

Amino acid	P-II-A-1-a P-II-B-1-a	P-II-A-1-b P-II-B-1-b	P-IIX-B-b
	Ratio of residues		
Asp	1.0	1.0	1.0
Thr	0.8	1.0	0.5
Ser	0.7	0.7	0.7
Glu	1.2		0.8
Gly	2.7	1.0	5.4
Ala	0.8		
½Cys	0.3		
Val	1.0	1.0	0.7
Ile	0.5	0.8	0.5
Leu	0.7		0.8
Tyr			0.6
Phe		0.6	

[a] Amino acid residues were normalized to Asp = 1.0. Values
for Thr and Ser are uncorrected for hydrolysis loss.

Table III

Hydroxamate Incorporation from the Incubation of Phenyl
Tetrahydrofurfuryl Sulfite with Modified Pepsin Species

Modified Reagent	Modified aspartate	Mol of hydroxamate found per mole of pepsin[a]		No. of bands on gel after PTFS-NH$_2$OH treatment
		Control	with PTFS	
1. Diazoacetyl-D,L-norleucine methyl ester	215	2.9	8.9	3
2. α-Diazo-p-bromo-acetophenone	probably 215	5.1	turbid	2[b]
3. 1,2-Epoxy-3-(p-nitrophenoxy)-propane	32	0	7.5	4
4. Reagents 2 and 3	32 and probably 215	5.8	13.6	4
5. Native			4.3	2[c]

[a] A 2.5 mol excess of PTFS was added 10 times at 30 min intervals.
Only CH$_3$CN was added to the control.

[b] One band was very diffuse.

[c] One major band and one minor band were observed as in the case
of untreated native pepsin.

Table IV

[θ] Values of Modified Pepsin Species at 215 and 290 nm

Modifying Reagent	$[\theta]_{215\ nm}$	$[\theta]_{290\ nm}$
1. Diazoacetyl-D,L-norleucine methyl ester	-10.5×10^3	+120 (at 285 nm)
2. α-Diazo-p-bromo-acetophenone	Turbid not measured	
3. 1,2-Epoxy-3-(p-nitro-phenoxy)propane	-9.2×10^3	+ 150
4. Reagents 2 and 3	$- 8.2 \times 10^3$	+ 78
5. Native pepsin with PTFS (4 mols hydroxamate found)	-9.8×10^3	+ 98
6. Native pepsin	-9.8×10^3	+ 70

[θ] is given as deg x cm^2 x $decimole^{-1}$. All pepsins were dissolved in 0.01 M sodium acetate, pH 4.0.

As shown in the third column of Table II, we also obtained a hydroxamate positive fragment which seemed to contain Asp_1, Thr_1, Ser_1,
Glu_1, Gly_5, Val_1, Ile_1, Leu_1 and Tyr_1. This fragment was the
largest peptide isolated in this trapping experiment. At the
present time we cannot unequivocally assign the position of this
fragment in the pepsin sequence. We plan to carry out Edman degradation on the fragment in the near future.

Purified pepsin was treated with several previously reported
active site modifying reagents (see the Methods section), and the
modified pepsins obtained were incubated with PTFS for ten 30-min
periods as described above. The results are summarized in Table III.
Hydroxamate was found in most of the modified pepsin species incubated
both with and without PTFS in the presence of 0.01 M hydroxylamine.
Considerably more hydroxamate incorporation in the presence of PTFS,
however, was observed with the modified pepsins than with the native
enzyme. These results required further study. According to Bornstein and Balian (31), bovine pancreatic ribonuclease was cleaved at
Asn-Gly and Asn-Leu linkages by treatment with 2 M hydroxylamine at
pH 9.0. If pepsin is susceptible to denaturation at pH 5.3 upon
chemical modification, similar cleavages might occur even upon a
mild treatment with 0.01 M hydroxylamine and could result in the incorporation of multiple hydroxamate groups. Pepsin contains two
Asn-Leu linkages (residues 37-38 and 139-140) and one Asn-Ile linkage
(231-232) in its amino acid sequence (28). After the modified pepsins
were incubated with PTFS, all preparations were subjected to gel
electrophoresis. Multiple bands appeared on the gels as indicated
in Table III, and the results suggested strongly that the modified
pepsin species were far more susceptible to cleavage by 0.01 M
hydroxylamine than was the native enzyme.

To examine the possible conformational changes due to the chemical modification of pepsin, CD spectra of the modified enzymes were
measured in the region between 205 nm to 320 nm. Typical CD spectra
are shown in Figure 1. The values of $[\theta]$ in the near uv region,
250 to 320 nm, showed changes compared to the native enzyme for all
modifications. The values of $[\theta]$ in the far uv region, particularly
at 215 nm where the β-structure of pepsin appears to be reflected
(32), changed approximately $\pm 1,000^0$ with chemical modification of
the enzyme with large esterifying groups. No significant change,
however, was observed when hydroxamate groups were incorporated by
treatment of native pepsin with PTFS in the presence of hydroxylamine.

DISCUSSION

The mechanism shown in Figure 2 has been proposed for the pepsin-
catalyzed hydrolysis of sulfite esters (14). This mechanism involves
the postulated formation of an anhydride intermediate (V). If such a
species is formed in the pepsin-catalyzed hydrolysis of sulfite esters,

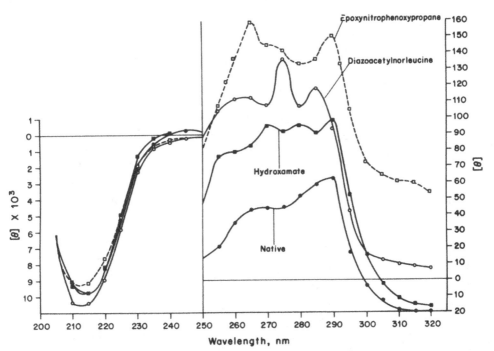

Figure 1. The results are expressed as mean residue ellipti-
cities in units of deg x cm^2 x decimole^{-1}. Protein samples
were dissolved in 0.01 M sodium acetate, pH 4.0, at the following concentrations: N-diazoacetyl-D, L-norleucine methyl ester-
modified pepsin, 0.244 mg/ml; 1,2-epoxy-3-(p-nitrophenoxy)-
propane-modified pepsin, 0.321 mg/ml; hydroxamate-containing
pepsin (4 mol/mole of pepsin), 0.227 mg/ml; native pepsin, 0.364
mg/ml. The light pathlength was 10 mm for the near ultraviolet
region (250-300 nm) and 1 mm for the far ultraviolet region
(205-250 nm). Each curve represents an average of three measurements.

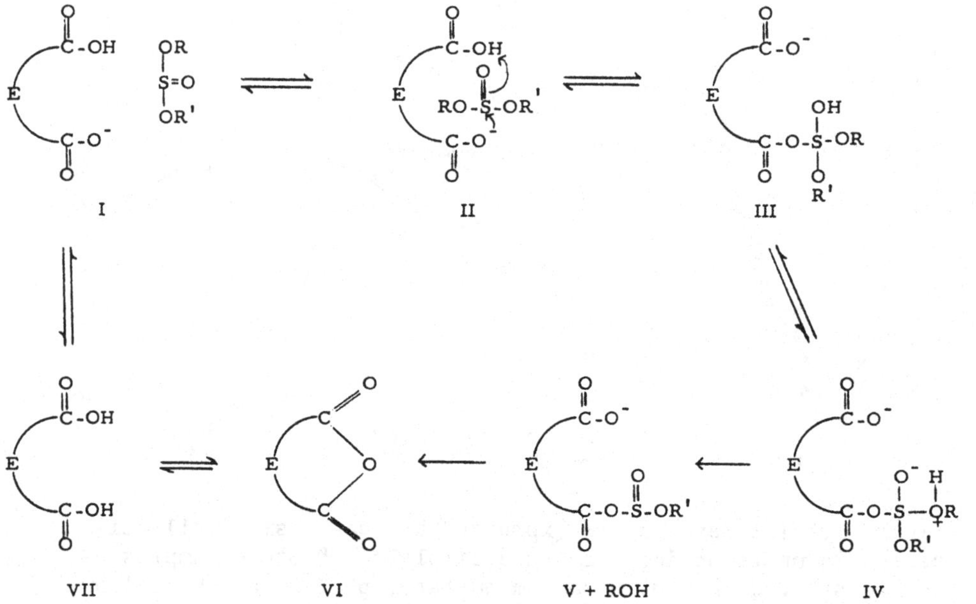

Figure 2.   Proposed mechanism for the pepsin-catalyzed
hydrolysis of sulfite esters.

a reactive nucleophile might trap the anhydride (V) is illustrated
in equation 1 below.

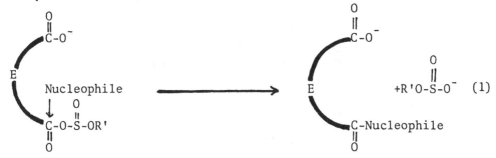

As shown in Table I, the incubation of pepsin with various
amounts of PTFS in the presence of 0.01 M hydroxylamine at pH 5.3
resulted in the incorporation of 1 to 4 mol of hydroxamate per mol
of pepsin with a concomitant partial loss of enzymatic activity.
These observations are consistent with the scheme of equation 1 in
which the potent nucleophile, hydroxylamine, attacks the mixed anhydride intermediate (V).

Many investigators have reported evidence from chemical modification studies with various esterifying reagents that there are at
least two catalytically essential β-carboxylates of aspartate residues
in the active site of pepsin (24,25,30,33). Additionally, Gross
and Morell (22) and Erlanger et al. (25) have demonstrated that an
aspartate β-carboxylate group which is not crucial for catalysis is
esterified by p-bromophenacyl bromide. It was, thus, of interest to
determine the nature of the carboxylate groups involved in the formation of enzyme-bound hydroxamate as a result of the treatment of pepsin with PTFS in the presence of hydroxylamine. Therefore, the
pepsin species containing the hydroxamate-modified residues were
subjected to dinitrophenylation and Lossen degradation, following
the procedure of Gross and Morell (22). The reaction sequence employed is shown in Figure 3.

As shown in Figure 3, if a glutamate residue is involved in
hydroxamate formation, the final product after Lossen rearrangement
and acid hydrolysis should be 2,4-diaminobutyric acid. If an
aspartate β-carboxyl group is trapped as a hydroxamate species,
2,3-diaminopropionic acid is expected as the final product. Amino
acid analysis showed only 2,3-diaminopropionic acid in the acid
hydrolyzate; thus, only aspartate residues were involved in the formation of the intermediates trapped with hydroxylamine during the
pepsin-catalyzed hydrolysis of a sulfite ester.

Since the nature of the groups involved in anhydride formation
in the pepsin-catalyzed hydrolysis of PTFS was elucidated, we
attempted to identify the particular residues taking part in catalysis.

$$
\begin{array}{l}
\quad\quad\quad\quad\quad\quad |\quad\quad\quad\quad\quad\quad\quad O \\
\quad\quad O\quad\quad C{=}O\quad\quad\quad\quad\quad\parallel \\
\quad\quad\parallel\quad\quad|\quad\quad\quad\quad\quad\quad\quad\parallel \\
\;\;\;\;\text{C-NH-C-CH}_2\text{-CH}_2\text{-C-OH (glutamate)} \\
E\Big\langle\quad\quad\;\; H \\
\quad\quad\quad\quad\quad\; H \\
\;\;\;\;\text{C-NH-C-CH}_2\text{-C-OH}\quad\quad\text{(aspartate)} \\
\quad\quad\parallel\quad\;|\quad\quad\quad\parallel\quad\;\parallel \\
\quad\quad O\quad\; C{=}O\quad\; O \\
\quad\quad\quad\quad\quad\; |
\end{array}
$$

$$\Big\downarrow\; \text{PTFS, NH}_2\text{OH}$$

$$
\begin{array}{l}
\quad\quad\quad\quad\quad\quad |\quad\quad\quad\quad\quad\quad\quad O \\
\quad\quad O\quad\quad C{=}O\quad\quad\quad\quad\quad\parallel \\
\quad\quad\parallel\quad\quad|\quad\quad\quad\quad\quad\quad\quad\parallel \\
\;\;\;\;\text{C-NH-C-CH}_2\text{-CH}_2\text{-C-OH} \\
E\Big\langle\quad\quad\;\; H \\
\quad\quad\quad\quad\quad\; H \\
\;\;\;\;\text{C-NH-C-CH}_2\text{-C-NHOH} \\
\quad\quad\parallel\quad\;|\quad\quad\quad\parallel \\
\quad\quad O\quad\; C{=}O\quad\; O \\
\quad\quad\quad\quad\quad\; |
\end{array}
$$

$$\Big\downarrow\; \text{F—C}_6\text{H}_3(\text{NO}_2)_2$$

$$
\begin{array}{l}
\quad\quad\quad\; O\quad\quad\quad H\quad\quad\quad O \\
\quad\quad\quad\; \parallel\quad\quad\;|\quad\quad\quad\parallel \\
\text{Pepsin-C-NH-C-CH}_2\text{-C-NH-O—C}_6\text{H}_3(\text{NO}_2)_2 \\
\quad\quad\quad\quad\quad\quad\; | \\
\quad\quad\quad\quad\quad\; C{=}O \\
\quad\quad\quad\quad\quad\quad\; |
\end{array}
$$

$$\Big\downarrow\; \text{Lossen rearrangement, OH}^-, \text{100}^\circ\text{C, 10 min}$$

$$
\begin{array}{l}
\quad\quad\quad\quad O\quad\quad\quad H \\
\quad\quad\quad\quad \parallel\quad\quad\;| \\
[\text{Pepsin-C-NH-C-CH}_2\text{-N=C=O}] \\
\quad\quad\quad\quad\quad\quad\; | \\
\quad\quad\quad\quad\quad\; C{=}O \\
\quad\quad\quad\quad\quad\quad\; |
\end{array}
$$

$$\Big\downarrow$$

$$
\begin{array}{l}
\quad\quad\quad\quad O\quad\quad\quad H \\
\quad\quad\quad\quad \parallel\quad\quad\;| \\
\text{Pepsin-C-NH-C-CH}_2\text{-NH}_2 + \text{CO}_2 \\
\quad\quad\quad\quad\quad\quad\; | \\
\quad\quad\quad\quad\quad\; C{=}O \\
\quad\quad\quad\quad\quad\quad\; |
\end{array}
$$

$$\Big\downarrow\; \text{HCl, 110}^\circ\text{C, 20 hr}$$

$$
\begin{array}{l}
\quad\quad\quad\quad\quad\quad\quad\quad\; \text{CH}_2\text{-NH}_2 \\
\quad\quad\quad\quad\quad\quad\quad\quad\; | \\
\text{Amino acids} + \text{H}_2\text{N-CH} \\
\quad\quad\quad\quad\quad\quad\quad\quad\; | \\
\quad\quad\quad\quad\quad\quad\quad\quad\; \text{CO}_2\text{H}
\end{array}
$$

Figure 3.   Degradation scheme for enzyme-bound hydroxamate.

A solution of pepsin containing approximately 3 mol of hydroxylamine per mol of enzyme was digested by pepsin and the undigested protein was separated on a column of Sephadex G-25 from the peptides resulting from digestion. The hydroxamate-positive peptides were purified further by two high-voltage electrophoreses and descending paper chromatography. This was followed by acid hydrolysis and amino acid analysis. Preliminary experiments indicated that the hydroxamate groups attached to the β-carboxylate of aspartate were stable at pH 2.0 overnight; however, drastic acidic conditions were avoided for the isolation and purification of the peptides. Two sets of the separated peptides showed identical amino acid compositions, and an apparently unique large peptide was also obtained and characterized. An additional peptide showing a hydroxamate-positive reaction was isolated, but this fragment contained a great variety of amino acids, and it was impossible to identify its location in the pepsin sequence. Thus, the composition of this fragment is not listed in the table.

One peptide (see first column of Table II) appeared to correspond to the sequence between Gly-208 and Leu-220 which includes Asp-215, previously identified as an active site residue by chemical modification with diazo amino acid derivatives. This peptide contained 0.3 cysteine residue according to our amino acid analysis results. It is well known that cysteine decomposes during acid hydrolysis and that the color factor for cysteine is one-half of those for other amino acids except proline. The low recovery of cysteine in the analysis of this peptide could be due to both factors. Considering the uncertaintities also in the serine and threonine analytical data, the assigned sequence surrounding Asp-215 for our hydroxamate-containing peptide might be moved a few residues to either side.

The sequence of the second peptide (see second column of Table II) which contained one phenylalanine residue was assigned as a fragment in the vicinity of Asp-32, which is considered to be an active site residue on the basis of chemical modification data obtained with 1,2-epoxy-3(p-nitrophenoxy)propane (26). Examining the amino acid sequence of pepsin in this region, from Val-29 to Ser-35, threonine and serine residues are present. It seems possible that this hydroxamate-positive fragment might contain several amino acid residues which undergo decomposition during acid hydrolysis. In this event, the peptide fragment might be a slightly longer chain than that reported here.

The third peptide (see third column of Table II) was a large one, composed of approximately thirteen amino acid residues according to our analytical data. The only unique residue was tyrosine, and it is difficult to assign the position of this fragment in the amino acid sequence of pepsin. Such an assignment must await the results of Edman degradation on the fragment.

Our observations that hydroxamate groups are incorporated at the Asp-32 and Asp-215 residues, and also at least one more Asp site, indicate that the β-carboxylate groups of all of these residues can function as attacking nucleophiles forming anhydride species in the pepsin-catalyzed hydrolysis of sulfite esters. The question which can be asked at this point is whether the β-carboxylates of Asp-32 and Asp-215 and another Asp residue are clustered together in one active site or whether pepsin might contain more than one distinct active site. The Asp-32 and Asp-215 residues (pepsin numbering) have been found to lie close together in the active site of penicillopepsin, according to findings reported on the x-ray structure of this enzyme (34). In view of the extensive homology between penicillopepsin and porcine pepsin, a similar situation must hold for pepsin. Inspection of the penicillopepsin structure did not reveal any additional Asp residue in extremely close proximity to Asp-32 and Asp-215. Considering the present uncertainty about the position(s) of hydroxamate incorporation other than at the Asp-32 and Asp-215 residues, it seems conceivable that for sulfite esterase action there is a second active site region. We must reserve judgment on the problem of whether, if such a site exists, it corresponds in the case of peptidase action to a second active site or possibly to a binding site.

In a recent article (35), evidence based on comparative hydrolysis and transpeptidation data was presented by Silver et al. to argue against the postulation of simple amino-enzyme intermediates in the reactions of pepsin with peptides. The suggestion was made that, for substrates which might be expected to form common simple amino-enzymes, conformational differences between the various pepsin-substrate complexes formed might account for the varying degrees of hydrolysis vs transpeptidation seen under identical experimental conditions. We would like to suggest instead the possibility that, as in the case of sulfite esters, the β-carboxylates of either Asp-32 or Asp-215 can function as the attacking nucleophiles. Peptide substrates possessing common amino-terminal groups but different acyl groups might then show preferential productive binding, allowing attack by either the β-carboxylate of Asp-32 or that of Asp-215 at the carbonyl moiety of the reactive peptide bond. If, under given experimental conditions, hydrolysis is facilitated at one carboxylate group and transpeptidation at the other, this could explain the many confusing data which have been reported for the interactions of pepsin with peptide substrates.

In addition to our hydroxylamine-trapping experiments performed with native pepsin and sulfite esters, pepsin modified by several known active site-directed esterifying reagents was incubated with PTFS in the presence of hydroxylamine. The degree of hydroxamate incorporation was substantially larger in the latter experiments than that seen in the case of the native enzyme.

Concomitantly, multiple bands appeared on gel electrophoresis performed on samples of esterified pepsins which had been treated with
PTFS in the presence of hydroxylamine. This seems to indicate that
the esterified pepsins were susceptible to peptide bond cleavage on
mild treatment with PTFS and 0.01 M hydroxylamine at pH 5.3. A
possible explanation for these observations came from the examination
of the CD spectra of modified pepsin species. All modifications
(including the introduction of hydroxamate groups in native pepsin)
gave altered spectra between 250 nm and 320 nm since new groups
were introduced into the pepsin molecule. The spectra below 250 nm
which reflect the backbone structure of the protein, and particularly
the negative trough at 215 nm, showed a change of approximately ±
$1,000^{\circ}$ in the $[\theta]$ value upon esterification of the active site Asp
groups. On the contrary, no change in this region was detected on
the introduction of 4 mol hydroxamate per mole of pepsin. These
results indicate that the structure of the pepsin molecule changed
appreciably when bulky esterifying groups were introduced at active
site residues, but not when hydroxamate groups were introduced in
native pepsin. Such structural changes may account for the height-
ened susceptibility of esterified pepsin to undergo peptide bond
cleavage on treatment with PTFS and hydroxylamine.

An important remaining problem which we are currently studying
is the determination of the reasons for the remaining catalytic
activity seen for pepsin in which three to four hydroxamate groups
have been introduced at the reactive Asp residues.

ACKNOWLEDGMENT

The partial support of this research by a grant from the
National Science Foundation and by a John Simon Guggenheim Memorial
Fellowship (ETK) is gratefully acknowledged.

SUMMARY

By the use of sulfite ester substrates together with hydroxylamine as a highly reactive trapping agent, we have been able to obtain
strong evidence for the intermediacy of enzyme-bound anhydride species
in the pepsin-catalyzed hydrolysis of these substrates. From our
observations that in the trapping experiments hydroxamate functions
are introduced at the β-carboxylates of Asp-32, Asp-215 and at least
one additional Asp residue, it appears that several reactive carboxylate species can function as nucleophiles against sulfite esters,
leading to the formation of anhydride species. Because the location
of the hydroxamate incorporated into pepsin other than at the Asp-32
and Asp-215 residues is unknown, it remains conceivable that, at
least for the action of pepsin on sulfite substrates, there are two
distinct active site regions. If the possibility is considered that

peptides possessing common amino-terminal residues but different acyl residues may bind productively in different fashions so that in some cases the β-carboxylate of Asp-32 acts as the attacking nucleophile while in others the β-carboxylate of Asp-215 acts in this way (as has been observed for sulfites), much of the confusion in the literature concerning the reactions of pepsin with peptidase may be explained.

REFERENCES

1. Clement, G. E. (1973) Prog. Bioorg. Chem. 2, 177-238
2. Knowles, J. R. (1970) Philos. Trans. R. Soc. London Ser.B., 257, 135-146
3. Kitson, T. M., and Knowles, J. R. (1971) Biochem. J. 122, 249-256
4. Newmark, A. K., and Knowles, J. R. (1975) J. Am. Chem. Soc. 97, 3557-3559
5. Antonov, V. K., Rumsh, L. D., and Tikhodeeva (1974) FEBS Lett. 46, 29-33
6. Silver, M. S., and Stoddard, M. (1975) Biochemistry 14, 614-621
7. Takahashi, M., and Hofmann, T. (1975) Biochem. J. 147, 549-263
8. Zeffren, E., and Kaiser, E. T. (1967) J. Am. Chem. Soc. 89, 4204-4208
9. Sachdev, G. P., and Fruton, J. S. (1975) Proc. Natl. Acad. Sci. U.S.A. 72, 3424-3427
10. Fruton, J. S. (1976) Adv. Enzymol. 44, 1-36
11. Fahrney, D., and Reid, T. (1967) J. Am. Chem. Soc. 89, 3941-3943
12. Zeffren, E., and Kaiser, E. T. (1968) Arch. Biochem. Biophys. 126, 965-967
13. May, S. W., and Kaiser, E. T. (1971) J. Am. Chem. Soc. 93, 5567-5572
14. May, S. W., and Kaiser, E. T. (1972) Biochemistry 11, 592-600
15. Chen, H. J., and Kaiser, E. T. (1974) J. Am. Chem. Soc. 96, 625-626
16. Cornish-Bowden, A. J., and Knowles, J. R. (1969) Biochem. J. 113, 353-362
17. King, L.-H., and Kaiser, E. T. (1974) J. Am. Chem. Soc. 96, 1410-1417
18. Hauser, C. R., and Renfrow, W. B., Jr. (1943) Org. Synth. Coll. 2, 607-609
19. O'Donnell, J. F., Ayres, J. T., and Mann, C. K. (1965) Anal. Chem. 37, 1161-1162
20. Chow, R. B., and Kassell, B. (1968) J. Biol. Chem. 243, 1718-1724
21. Bergmann, F., and Segal, R. (1956) Biochem. J. 62, 542-546
22. Gross, E., and Morell, J. L. (1966) J. Biol. Chem. 241, 3638-3639
23. Moore, S., and Stein, W. H. (1963) Methods Enzymol. 6, 819-831

24. Rajagopalan, T. G., Stein, W. H., and Moore, S. (1966) J. Biol. Chem. 241, 4295-4297

25. Erlanger, B. F., Vratsanos, S. M., Wasserman, N., and Cooper, A. G. (1967) Biochem. Biophys. Res. Commun. 28, 203-208

26. Hartsuck, J. A., and Tang, J. (1972) J. Biol. Chem. 247, 2575-2580

27. Tang, J., Sepulveda, P., Marciniszyn, J., Jr., Chen, K. C. S., Huang, W.-Y., Too, N., Liu, D., and Lanier, J. P. (1973) Proc. Natl. Acad. Sci. U.S.A. 70, 3437-3439

28. Sepulveda, P., Marciniszyn, J., Jr., Liu, D., and Tang, J. (1975) J. Biol. Chem. 250, 5082-5088

29. Rajagopalan, T. G., Moore, S., and Stein, W. H. (1966) J. Biol. Chem. 241, 4940-4950

30. Nakagawa, Y., King Sun, L.-H., and Kaiser, E. T. (1976) J. Am. Chem. Soc. 98, 1616-1617

31. Bornstein, P., and Balian, G. (1970) J. Biol. Chem. 245, 4854-4856

32. Perlmann, G. E. (1967) "Ordered Fluids and Liquid Crystals", Adv. in Chemistry Series No. 63, pp. 268 (Amer. Chem. Soc.)

33. Paterson, A. F., and Knowles, J. R. (1972) Eur. J. Biochem. 31, 510-517, and references therein

34. Hsu, I. N., Delbaere, L. T. J., James, M. N. G., Hofmann, T., "The Crystal Structure of Penicillopepsin at 2.8 A Resolution" (1976) Conference on Acid Proteases, Norman, Oklahoma, November 21-24 (Chapter 5 in this volume)

35. Silver, M. S., Stoddard, M., and Kelleher, M. H. (1976) J. Am. Chem. Soc. 98, 6684-6690

NEW DATA ON PEPSIN MECHANISM AND SPECIFICITY

Vladimir K. Antonov

M. M. Sheymakin Institute of Bioorganic Chemistry
USSR Academy of Sciences, Moscow

Understanding of the detailed catalytic mechanism of porcine pepsin requires knowledge of the three-dimensional structure of its active site. Currently, the results of x-ray crystallographic studies of this enzyme are still preliminary (1,2). The study of specificity and mechanism is, therefore, particularly useful in formulating the interpretation of the catalytic functions of the structural features in the crystallographic models.

Studies on pepsin specificity (3) have shown that the enzyme has primary affinity toward the two side chains of the amino acid residues adjacent to the substrates cleaving peptide bond. But the kinetic data indicated some "secondary interactions" of several additional residues on each side of the hydrolyzed bond. Therefore, the dimensions of the active site must be large enough to accomodate all these interactions. From this information, two interesting questions arose: (a) What is the total length of the substrate binding region of pepsin, and (b) What are the relative binding preferences of substrate side chains at the different binding loci within the pepsin binding domain?

To answer these questions, we have analyzed the frequency of appearance of various amino acid residues at the loci near the sessile bond in protein substrates (4). This was statistically analyzed for more than 500 amino acid sequences in which the peptic hydrolytic sites are known. The calculated specificity indices, S_{ij}, for some of the residues (i) and positions (j) are shown in Table I. This index is a measure of the preferability, undesirability, or indifference for a given amino acid side chain. The calculation included ten residues on each side of the hydrolyzed bond. We also

Table I

Specificity Indices (S_{ij})* for Protein Cleavage by Pepsin

Positions in substrate sequence									
P_3		P_2		P_1		P_1'		P_2'	
Ser	+1.53	Asn	+1.70	Leu	+2.46	Trp	+1.52	Arg	+1.45
Pro	+1.45	Glu	+1.63	Phe	+2.28	Tyr	+1.52	Thr	+1.36
Gly	+1.45	Ala	+1.36	Trp	+1.67	Ile	+1.40	Ile	+1.34
Thr	+1.29	Val	+1.24	Glu	+1.49	Phe	+1.39	Ala	+1.33
Trp	-1.70	Trp	-1.27	Thr	-1.32	Asn	-1.25	Met	-1.23
Arg	-1.72	Lys	-1.32	Gly	-1.52	Met	-1.29	His	-1.33
His	-1.78	Phe	-1.43	His	-1.66	Arg	-1.37	Tyr	-1.37
Phe	-2.03	Pro	-1.43	Val	-1.95	Ser	-1.37	Asp	-1.69

*Calculated according to formulas: $S_{ij} = k_{ij}/[1+\Delta X_i]$ for $k_{ij} \geqslant 1$ and $S_{ij} = [1 - \Delta X_i]/k_{ij}$ for $k_{ij} < 1$; $k_{ij} = 20 b_{ij}/a_i \sum_{i=1}^{20} \frac{b_{ij}}{a_i}$ and $\Delta X_i = 2\sqrt{\sum_{i=1}^{20} (\bar{k}_{ij}-k_{ij})^2/[a_i-1]}$ where b_{ij} is a number of amino acids of a given type (i) in a given position (j); a_i is a total number of amino acids of a given type in all sequences studied; $\bar{k}_{ij}$ is the mean value of k_{ij}.

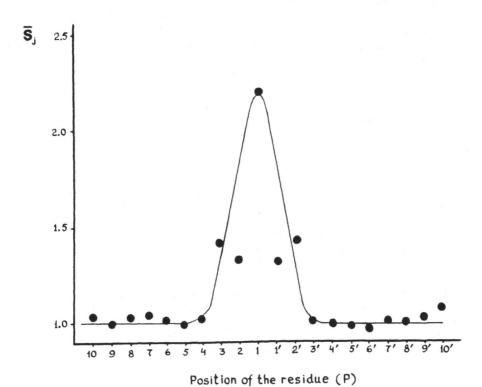

Position of the residue (P)

Figure 1.   Dependence of the mean specificity index $\bar{S}j$ on the position of residue (P) in the polypeptide chain of the substrate.   The bond to be hydrolyzed is between positions P_1 and P_1';

$$\bar{S}_{j.} = \sum_{i=1}^{20} S_{ij}/20,$$

where S_{ij} is the specificity index for a given amino acid (i) in a given position (j) (4).

stimated the mean deviation from casual distribution of residues
for each position, which characterized the "rigidity" of requirement
imposed on the structure of an amino acid side chain in a given
binding position.

We have found that these deviations are greatest in locus S_1,
noticeable enough in loci S_2, S_3, S_1', and S_2', and differ but slightly
from unity for other positions (Fig.1). Therefore, the active site
of pepsin is adjusted to accomodate specifically at least five amino
acid residues: three toward the NH_2-terminal from the sensitive bond
and two toward the COOH-terminal. The most strict requirements are
imposed on the structure of the side chain located in the locus S_1.
The narrow specificity requirements at this position are in contrast
to the broader specificity at position S_1', which can accomodate side
chains according to the degree of hydrophobicity. Powers et al. have
also obtained similar results (see Chapter 9 in this volume).

The data presented so far provides no information about the
relative rate at which pepsin hydrolyzes various peptide bonds. It
would be important, therefore, to compare these results with the
hydrolytic kinetics of some of their bonds in low-molecular-weight
peptide substrates. Despite the fact that much information has been
obtained in kinetic studies of pepsin using synthetic substrates (3),
there always seems to be a need for new highly soluble and easily
available peptide substrates of pepsin. We have synthesized and
studied a new type of pepsin substrate (Table II) whose solubility
was improved as a result of the introduction into the COOH-termini
of these peptides an γ-aminopropylmorpholine group. The p-nitro-
phenylalanine residue can be placed into positions P_1 (6), and P_1'
of the substrate, allowing the kinetics to be studied spectrophoto-
metrically.

Data on kinetic parameters which we obtained for hydrolysis of
a number of di-, tri-, and tetrapeptides (Table II), as well as the
data from other workers (3,6-9), are in qualitative agreement with
the results of statistical analysis of the peptide specificity on
protein substrates described above. It should be mentioned here
that the changes in the substrate structure affect mainly the catalytic constants, k_{cat}, rather than the affinity between the substrate and the enzyme as reflected in the values of K_m (see also 9,10).

The effect of the side chains of substrate residues not directly
adjacent to the sensitive bond has been demonstrated in the following
experiments. It was shown that, in the substrate Ala-Phe(NO$_2$)-Ala-
APM, the bond Phe(NO$_2$)-Ala was only very slowly hydrolyzed by pepsin;
however, when the tripeptide Z-Leu-Ala-Ser-OH, which is not susceptible to enzymatic hydrolysis, was included in the hydrolysis solution,
the rate of the substrate hydrolysis increased sharply (k_{cat} = 0.48
min^{-1}, K_m=2.86 mM). Now, only the bond Ala-Phe(NO$_2$) was hydrolyzed.

Table II

Kinetic Constants of Pepsin Catalyzed Hydrolysis of
Synthetic Substrates

Conditions: 0.1 M acetate buffer, pH 4.0, 25°C, 5%(vol.)DMF

No.	Substrate	K_{cat}	K_m	K_{cat}/K
		min^{-1}	mM	$min^{-1} mM^{-1}$
1	Z-Phe(NO$_2$)-Phe-APM[a]	2.15	0.69	3.12
2	Ac-Phe(NO$_2$)-Phe-APM	1.2	0.3	3.5
3	Z-Phe(NO$_2$)-Gly-APM	Is not hydrolyzed		
4	Z-Phe(NO$_2$)-Ala-APM	0.7	0.93	0.75
5	Z-Phe(NO$_2$)-Val-APM	0.44	0.77	0.57
6	Z-Phe-Phe(NO$_2$)-APM	1.05	0.44	2.4
7	Z-Ala-Phe(NO$_2$)-APM [b]	0.125	0.54	0.08
8	Z-Leu-Phe(NO$_2$)-APM	0.72	1.33	0.54
9	Z-Val-Phe(NO$_2$)-Phe-APM	18.3	0.16	115
10	Z-Val-Phe(NO$_2$)-Ala-APM	32.2	1.0	32.2
11	Z-Phe(NO$_2$)-Phe-Arg-OMe	2.12	0.95	2.23
12	H-Gly-Gly-Phe-Phe-APM[b]	57.0	3.9	15.0
13	Z-Leu-Val-Phe(NO$_2$)-Ala-APM	262	0.35	750
14	Z-Ala-Ala-Phe(NO$_2$)-Phe-APM	2630	1.51	1740

[a]APM = NH(CH$_2$)$_3$N◯O .
[b]Without DMF.

These data can be interpreted by assuming that, when the substrate (Ala-Phe(NO$_2$)-Ala-APM) alone was being bound to pepsin, the
side chain of Phe(NO$_2$) settled in the most specific locus S_1.   In
the presence of the activating tripeptide, the side chains of
Ser, Ala, and Leu occupied preferentially the binding loci S_2, S_3,
and S_4, in the same order.   As a result, the substrate can now be
bound only in such a way that Phe(NO$_2$) resides in the second hydrophobic locus S_1', and the bond Ala-Phe(NO$_2$) is hydrolyzed.   As suggested from the values of specificity indices (4), this binding
arrangement is most preferable as compared to any alternative combination of the substrate and activating tripeptide bindings (Fig.2).

Apart from the interaction between the side chains of the polypeptide substrate and the enzyme, there may be other interactions
important for the efficiency of peptic catalysis.   As was shown by
Fruton et al. (11), if the methylene group in residue P_1' is sub-
stituted for the carbonyl group, the substrate becomes an inhibitor.
We found (12) that substitution of the -NH- group is residue P_1 by
an isosteric oxygen atom also resulted in a resistance to peptic
hydrolysis.   This compound was competitive inhibitor with $K_i \approx K_m$ of
the corresponding substrate (Table III).   Therefore, formation of
hydrogen bonds to the dipeptide region of the substrate is a prerequisite of peptic catalysis.

The very next important question is: What is the orientation of
the substrate in the enzyme's active site relative to the binding
loci? To solve this problem kinetically, it is necessary to find a
compound which interacts preferentially with one of the loci S_1 or S_1'.
We have found that ethanol acts as a competitive inhibitor of hydrolysis of Z-Phe(NO$_2$)-Ala-APM and as a non-competitive inhibitor of
hydrolysis of Z-Ala-Phe(NO$_2$)-APM (Fig.3).   This may be explained by
assuming that ethanol is bound in locus S_1 with the formation of a
ternary enzyme-ethanol-Z-Ala-Phe(NO$_2$)-APM complex (Fig.4).   The
possibility that ethanol interacts with some other place on the
enzyme, changing the conformation of the active site, is rather improbable due to the fact that this solvent, at concentration studied,
does not inhibit the hydrolysis of Z-Phe(NO$_2$)-Phe-APM.

The pepsin crystals used in the x-ray crystallographic studies
are prepared at the presence of ethanol (1,2).   It is quite possible
that ethanol molecules can eventually be identified on the three-
dimensional structure of pepsin crystals to confirm the kinetic
studies.

Concerning the mechanism of pepsin action itself, it is expe-
dient to focus on three problems: (1) what is the cause of the
known pH-dependence of pepsin catalysis; (2) are there amino-enzyme
and acyl-enzyme intermediates in the pathway of pepsin action; and
(3) what is the chemical nature of the intermediates.

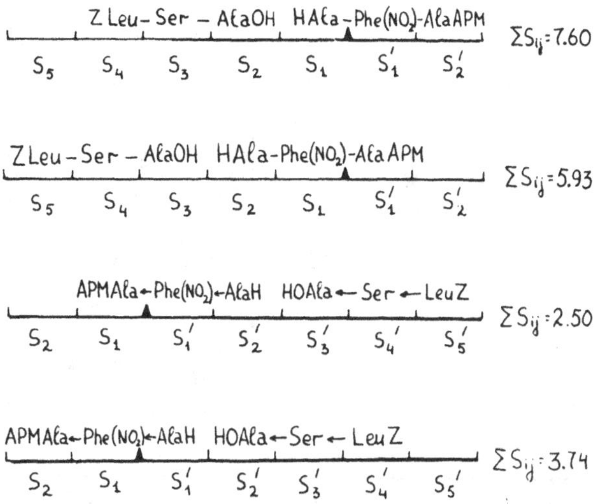

Figure 2. Possible ways of binding substrate H-Ala-Phe(NO$_2$)-Ala-APM and activator Z-Leu-Ser-Ala-OH in the active site of pepsin. ΣS_{ij} is the sum of specificity indices (4).

Table III

Interaction of the Dipeptide Substrate and its Isosteric Analogues with Pepsin

Conditions: 0.1 M acetate buffer, pH 4.0, 37°C 6%(vol) CH$_3$OH; $|E|_o$ = 2.9 10^{-5} M

Compound*	K_{cat}	K_m or K_i
	sec^{-1}	mM
Ac-L-Phe-L-Phe-OMe	0.1	1.90
Ac-L-Pla-L-Phe-OMe	-	-
Ac-D-Pla-L-Phe-OMe	-	-

*Pla- Phenyllactyl.

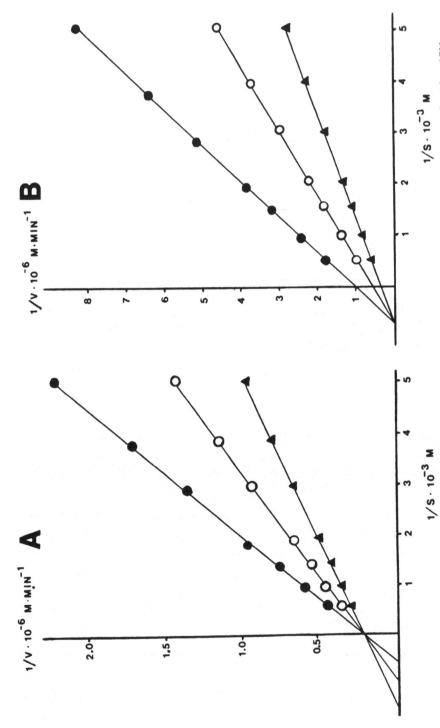

Figure 3. Inhibition by ethanol of pepsin catalyzed hydrolysis of: (A) Z-Phe(NO$_2$)-Ala-APM and (B) Z-Ala-Phe(NO$_2$)-APM. Conditions: 0.1 M acetate buffer, pH 4.0, 25°C. Ethanol concentrations (in vol.%) : (▲) - 0, (O)-2.5 and (●) - 5.

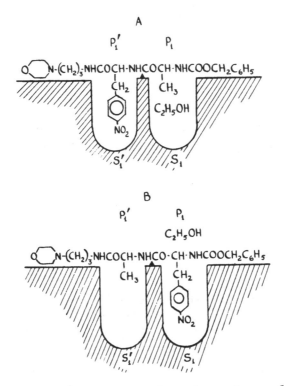

Figure 4. Schematic presentation of complexes formed by pepsin with ethanol and substrates, (A): Z-Ala-Phe(NO$_2$)-APM, (B): Z-Phe(NO$_2$)-Ala-APM.

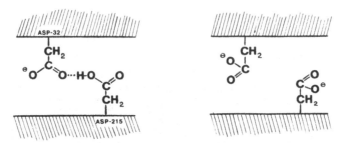

Figure 5. Possible mode of localization of Asp-32 and
Asp-215 carboxyls in the pepsin active site.

A. $E\text{-COOH} + RCONHR^1 \rightleftharpoons E\text{-CO-O-COR} \xrightarrow{R^2NH_2} E\text{-COOH} + RCONHR^2$

$\downarrow H_2O$

$RCOOH + E\text{-COOH}$

B. $E\text{-COOH} + RCONHR^1 \rightleftharpoons E\text{-CONHR}^1 \xrightarrow{R^2COOH} E\text{-COOH} + R^2CONHR^1$

$\downarrow H_2O$

$R^1NH_2 + E\text{-COOH}$

Figure 6. Pepsin-catalyzed acyl (A) and amino (B) transfer.

From the pH-dependence data (10), the pK_a values of the catalytically essential carboxyl groups of Asp-32 and Asp-215 are about
1.5 and 4.5, respectively. This difference can be attributed (13)
to the existence of a hydrogen-bonded system between these two
carboxylic groups. Once a proton has been eliminated from the system,
the orientation of the carboxyl groups may change, and as a consequence, the catalytic activity decreases (Fig.5). Such an interpretation is consistent with the pH-dependence of pepsin catalysis,
provided the carboxylate ion of Asp-32 is involved in the catalysis
at the rate-limiting step of the process.

The pepsin-catalyzed transpeptidation reactions, both the amino
transfer (14) and the acyl transfer types (15) (Fig.6), suggest the
formation of at least two intermediates in which fragments of the
substrate are bound to the enzyme. For a long time, however, no
transpeptidation reaction of the amino transfer type could be found
for substrates containing a blocked COOH-terminal carboxyl group
adjacent to the bond being cleaved. The existence of the "amino-
enzyme" in the pepsin hydrolysis pathway of conventional substrates
was therefore somewhat doubtful (16). As for the transpeptidation
of the acyl transfer type, it has been found so far only on substrates with a free NH_2-terminal amino group.

We have developed a technique for measuring initial formation
rates of the products resulting from the amino transfer type of
transpeptidation (17,18). To accomplish this, we used a chromophore
acceptor which changed its spectral characteristics once an amide
bond has been formed. The transpeptidation rate constant, k_5, can
be determined by analyzing the dependence of the initial rates on
the acceptor concentration (Fig.7).

In our studies, transpeptidation constants were compared for two
substrate pairs containing the same amino acid residue to be transferred (Table IV). From this data, a common intermediate product,
"amino-enzyme", was shown to exist for each substrate pair. Furthermore, the transpeptidation rate constants of substrates with a pro-
tected COOH-terminal amino acid were almost the same as that of
substrates with a free carboxyl group. Difficulties encountered in
detecting the transpeptidation product in the former case are due to
an unfavorable ratio between the rates of hydrolysis and amino transfer. The effect of pH on the transpeptidation rate constants was
studied (Fig.8); this constant was independent of pH in the pH range
from 3.7 to 5.6.

The facts of amino and acyl transfer in the course of peptic
catalysis, however, do not by themselves prove the covalent nature
of the intermediates, such as the acylation of the amion component
of the substrate by the enzyme carboxyls (amino-enzyme) or the anhydride intermediates formed by the acyl component of the substrates
by the enzyme carboxyls (acyl-enzyme) (Fig.6).

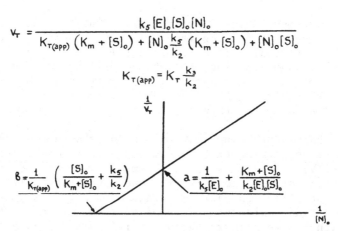

$$E + S \underset{}{\overset{K_S}{\rightleftharpoons}} ES \xrightarrow{k_2} E\text{-}B \xrightarrow{k_3} E + B$$
$$+$$
$$A$$

$$E\text{-}B + N \overset{K_T}{\rightleftharpoons} E\text{-}B\text{-}N \xrightarrow{k_5} E + BN$$

E - enzyme, **S** - substrate, **N** - acceptor,

A and **B** - products of hydrolysis, **BN** - product

of transpeptidation

$$V_T = \frac{k_5 [E]_o [S]_o [N]_o}{K_{T(app)} (K_m + [S]_o) + [N]_o \frac{k_5}{k_2} (K_m + [S]_o) + [N]_o [S]_o}$$

$$K_{T(app)} = K_T \frac{k_3}{k_2}$$

$$B = \frac{1}{K_{T(app)}} \left(\frac{[S]_o}{K_m + [S]_o} + \frac{k_5}{k_2} \right) \qquad a = \frac{1}{k_5 [E]_o} + \frac{K_m + [S]_o}{k_2 [E]_o [S]_o}$$

Figure 7. Kinetic Scheme for the reaction of transpeptidation
of the amino transfer type, and theoretical dependence of the
reciprocal transpeptidation rate on the reciprocal of the
acceptor concentration.

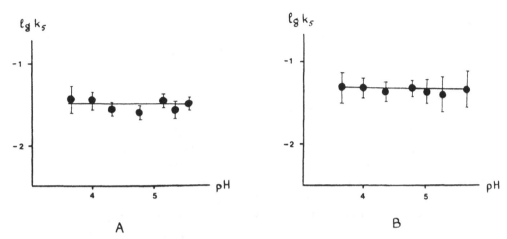

Figure 8. pH-dependence of the transpeptidation rate constant for (A) Ac-Phe-Tyr-OH and (B) Ac-Phe-Phe-APM. Acceptor - Z-Phe(NO$_2$)-OH.

Table IV

Transpeptidation Constants and the Ratio of the Hydrolysis and Transpeptidation Rates

Conditions: 0.1 M acetate buffer, pH 4.61, 37°C, Acceptor (A): Z-Phe(NO$_2$)-OH

Substrate	$k_5 \times 10^2$	$K_{T(app)}$	$V_{hyd}/V_{trans.}$
	min^{-1}	mM	$[A]_o$ 0.667 M
Ac-Phe-Tyr-OH	3.10± 0.4	0.16±0.04	22.0
Ac-Tyr-Tyr-OH	2.93± 0.75	0.42±0.18	22.2
Ac-Phe-Phe-APM	4.65± 0.54	1.70±0.54	351
H-Gly-Gly-Phe-Phe-APM	4.89± 0.09	0.18±0.12	380

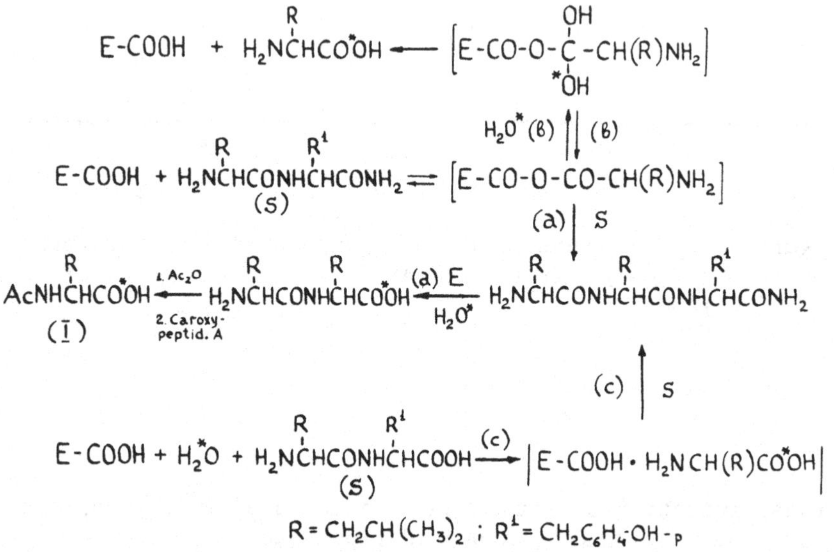

Figure 9. Possible means of $H_2^{18}O$ incorporation into the product of pepsin-catalyzed transpeptidation reaction of the acyl transfer type. Theoretical values of ^{18}O content in mg per 100 mg sample of (I) in (a): 0; (b): 0 ÷ 5.2; (c): 2.6. ^{18}O content in water is at 50%.

These facts cannot exclude the possibility that transpeptidation reactions involve non-covalent enzyme and substrate fragments. To clarify this, we studied the incorporation of ^{18}O atoms in both the enzyme and the substrate during the peptic catalysis. Two series of experiments have been conducted. In the first (Fig.9), a technique of activation analysis (19) was developed to determine the content of ^{18}O in the product of acyl transpeptidation formed upon incubation of (^{14}C) Leu-Tyr-NH$_2$ with pepsin in $H_2^{18}O$. Then one might expect: (a) the absence of ^{18}O incorporation if an acyl-enzyme of the anhydride type is involved; (b) the incorporation of various amounts of ^{18}O, depending on the relative rates of the intermediate anhydride to undergo reversible hydration; (c) the incorporation of one oxygen atom, if the cleavage of peptide bond involves water and the formation of a non-covalent complex of the enzyme and the amino component of the substrate.

We have found that ^{18}O is indeed incorporated into the product of acyl transpeptidation less than one atom per molecule (Fig.10). This evidence alone is insufficient to distinguish between the mechanism of hydrated anhydride (b) and the mechanism involving direct participation of water (c).

In the second series of experiments (Fig.11), we studied the ability of the carboxyl groups of Asp-32 and Asp-215 in the active site to exchange their oxygen with the oxygen isotope of water. If during the substrate hydrolysis, the covalent intermediates, anhydride and/or amino-enzyme, are formed and subsequently cleaved resulting from the attack of the enzyme carbonyls by water, then the carboxyl groups of the active site should exchange their oxygen with the oxygen of water. We incubated the substrate with pepsin in $H_2^{18}O$, then separated the enzyme, and treated it sequencially with two labeled inhibitors: epoxide which reacts with Asp-32 (20) and diazoketone which reacts with Asp-215 (21). The enzyme preparations treated with the inhibitors were isolated and hydrolyzed under mild alkaline conditions. The products obtained after cleavage of the inhibitors were subjected to ^{18}O analysis. In no case did the content of ^{18}O in the products exceed that of natural abundance (Fig.10). The following conclusions can therefore be made: (a) If the "acyl-enzyme" is formed, it is attacked by water on the carbonyl group of the acyl moiety of the substrate (this conclusion is consistent with the data on the incorporation of ^{18}O in the product of acyl transpeptidation); (b) the enzyme does not produce an amide intermediate which consists of one of the carboxyl groups (Asp-32 or Asp-215) and the amino component of the substrate.

Three interpretations of these unexpected results are possible: (A) The amide intermediate involves a carboxyl group other than that of Asp-32 and Asp-215; (B) the amino-enzyme is a non-covalent complex of the amino component of the substrate and pepsin; and (C) the

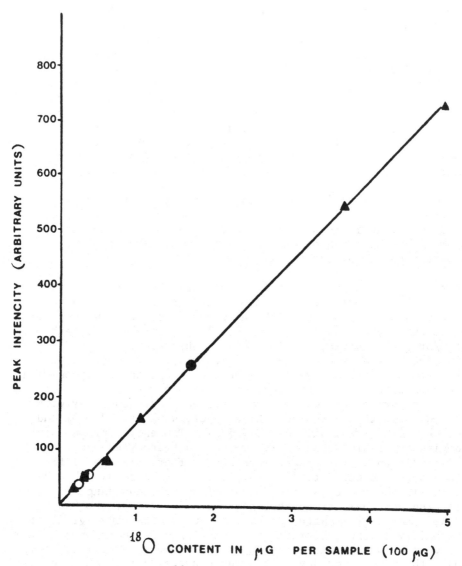

Figure 10.   Results of ^{18}O exchange experiments (▲)-calibration
Standards, (●) - Ac-Leu-OH (see Fig.9), (O) - $HOCH_2CONHCH$-
($CH_2C_6H_5$)COOH and (◑) - $HOCH_2CH(OH)CH_2C_6H_5$ (see Fig.11).

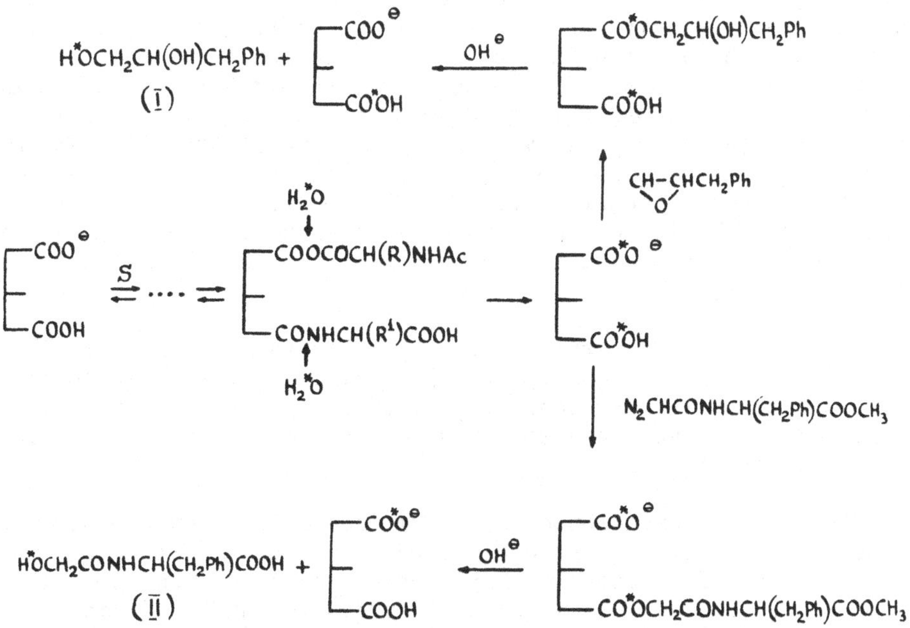

Figure 11. Schematic of experiments in which the ability of pepsin to exchange oxygen atoms of carboxyls in Asp-32 and Asp-215 with ^{18}O was studied. Theoretical values of ^{18}O content per 100 mg samples of (I) - 5.9: of (II) - 4.0 mg.

catalytically active carboxyl groups and the amino component to be
transferred are linked not by an amide bond, but by another type of
covalent bond.

As can be seen from the x-ray data available (see Chapters 2-5
in this volume) there are no other carboxyl groups located in the
vicinity of Asp-32 and Asp-215 residues, and it is unlikely that
any groups other than these two are involved in the catalysis.

The formation of the non-covalent amino-enzyme is difficult to
reconcile in considering the energetics of transpeptidation. The
free energy of amide bond formation (in the product of transpeptidation) is negative only for unionized forms of reagents (22). In
other words, the amino group of the amino component of the substrate
in a non-covalent complex with the enzyme should have pK_a of about
3-4 units lower than that in an aqueous solution. According to the
x-ray model, the only positively charged group of Arg-316 which may
perturb pK_a of the amino grouping is situated far enough from the
catalytic center of pepsin. It is still possible, however, to
explain the stability of the non-covalent amino-enzyme by a conformational change in the active site of pepsin due to substrate binding
(3).

The formation of a relatively stable tetrahedral intermediate
product involving the carboxyl of the enzyme and the amino component
of the substrate is also consistent with the absence of ^{18}O exchange.
In this case, however, the mechanism of its formation and its stability has to be explained.

From the aforementioned data, some conclusions on the structure
of the pepsin active site can be drawn. The active site seems to
be an extended cleft where carboxyl groups of Asp-32 and Asp-215
residues are located. These groups divide the substrate binding
into two parts which are able to accomodate specifically three
amino acid residues from the amino-terminal side and two residues
from the carboxyl-terminal side of the substrate. The orientation
of the substrate molecule is such that the amino-terminal residue of
the dipeptide unit is bound at the same locus as an ethanol molecule.
The active-site carboxyl groups, Asp-32 and Asp-215, are presumably
close to one another, forming a hydrogen-bonded system. The x-ray
data are consistent with these conclusions.

Further physico-chemical and x-ray analysis studies of pepsin
will furnish more information which will help to elucidate in detail
the mechanism of action of this enzyme as well as similar enzymes
belonging to the group of carboxyl proteinases.

ACKNOWLEDGMENT

The author gratefully acknowledges the contributions of L. M. Ginodman, L. V. Kozlov, T. N. Barshevskaya, Yu. V. Kapitannikov, A. G. Michailova, L. D. Rumsh, and A. A. Zinchenko and the kind help of N. S. Andreeva in analysis of the x-ray data.

SUMMARY

New data on the specificity and mechanism of action of porcine pepsin are presented, including statistical analysis of protein cleavage by the enzyme, kinetics of synthetic substrates, enzyme inhibition and activation, kinetics of transpeptidation reaction, and ^{18}O exchange studies.

From these data it was concluded that pepsin has an extended active site being able to accomodate specifically five amino acid residues of the substrate. The orientation of the substrate molecule relative to the ethanol binding loci in pepsin crystals has been determined. Pepsin mechanism includes "amino-enzyme" formation which chemically is not an amide, formed by the enzyme carboxyl with the amino fragment of the substrate.

REFERENCES

1. Andreeva, N. S., Fedorov, A. A., Guschina, A. E., Shutzkever, N. E., Riskulov, R. R., and Volnova, T. V. (1976) Doklady Akad. Nauk SSSR, 228, 480-483
2. Andreeva, N. S., Fedorov, A. A., Guschina, A. E., Shutzkever, N. E., Risculov, R. R., and Volnova, T. V. (1976) in Acid Proteases, Structure, Function and Biology (Tang, J., ed) pp. 000, (Chapter 2 in this volume), Plenum, New York
3. Fruton, J. S. (1976) in Advance Enzymol. 44, 1-35
4. Zinchenko, A. A., Rumsh, L. D., and Antonov, V. K. (1976) Bioorganicheskaiya Khimiya 2, 803-810
5. Tikhodeeva, A. G., Zinchenko, A. A., Rumsh, L. D., and Antonov, V. K. (1974) Doklady Akad. Nauk SSSR 214, 355-359
6. Inouye, K., and Fruton, J. S. (1967) Biochemistry 6, 1765-1777
7. Medzihradszky, K., Voynick, I. M., Medzihradszky-Schweiger, H., and Fruton, J. S. (1970) Biochemistry 9, 1154-1160
8. Sachdev, G. P., and Fruton, J. S. (1969) Biochemistry 8, 4231-4238
9. Fruton, J. S. (1970) in Advance Enzymol. 33, 401-443
10. Clement, G. E. (1973) in Progr. Bioorg. Mech. 2, 177-238
11. Humphreys, R. E., and Fruton, J. S. (1968) Proc. Natl. Acad. Sci. U.S.A. 59, 519-525
12. Rumsh, L. D., Tikhodeeva, A. G., and Antonov, V. K. (1974) Biokhimiya 39, 899-902

13. King, J. R. (1963) in Elucidation of Structures by Physical and Chemical Methods (Weissberger, A. ed) Russian edition, Vol. I, pp. 383, Khimiya, 1967, Moscow
14. Fruton, J. S., Fujii, S., and Knappenberger, M. H. (1961) Proc. Natl. Acad. Sci. USA 47, 759-765
15. Takahashi, M., and Hofmann, T. (1972) Biochem. J. 127, 35
16. Silver, M. S., and Stoddard, M. (1972) Biochemistry 11, 191-200
17. Antonov, V. K., Rumsh, L. D., and Tikhodeeva, A. G. (1974) FEBS Lett. 46, 29-33
18. Tikhodeeva, A. G., Rumsh, L. D., and Antonov, V. K. (1975)
19. Firsov, L. M., Khokhlacheva, M. A., and Gusinskii, G. M. (1976) Biokhimiya 41, 1176-1180
20. Tang, J. (1971) J. Biol. Chem. 246, 4510-4517
21. Kozlov, L. V., Ginodman, L. M., and Orekhovich, V. N. (1967) Biokhimiya 32, 1011-1019
22. Kozlov, L. V., Ginodman, L. M., Orekhovich, V. M., and Valueva, T. A. (1966) Biokhimiya 31, 315-321

PEPSTATIN INHIBITION MECHANISM

Joseph Marciniszyn, Jr., Jean A. Hartsuck, and Jordan Tang

From the Laboratory of Protein Studies
Oklahoma Medical Research Foundation, and
the Department of Biochemistry and Molecular Biology
University of Oklahoma School of Medicine
Oklahoma City, Oklahoma 73104

Pepstatin is a strong inhibitor for all acid proteases. It does not inhibit other groups of proteases, such as the neutral and alkaline proteases (1). The unusual potency of pepstatin toward acid proteases is indicated by its K_i which was reported by Kunimoto et al. (2) to be about 1×10^{-10}M for porcine pepsin. Although its chemical structure has been shown (3) to be a hexapeptide which contains two residues of an unusual amino acid, 4-amino-3-hydroxy-6-methylhepatanoic acid (statine), the mode of inhibition by pepstatin is unknown.

When the structure of pepstatin, which is isovaleryl-L-valy-L-valy-stayl-alanyl-statine, has been modified by esterification of its COOH-terminus, there is no effect on its inhibitory activity. Acetylation of its hydroxyl groups, however, drastically reduces the inhibition by pepstatin (4). Aoyagi et al. (5) have shown that three pepstatin fragments are weak inhibitors of pepsin. These data, together with the information gleaned from the use of two types of active-site directed irreversible inhibitors (6-9), suggest a common mechanism for the pepstatin inhibition of acid proteases.

EXPERIMENTAL RESULTS AND DISCUSSION

Inhibition of N-acetyl-statin and its Dipeptides

Since pepstatin is an extremely potent inhibitor, the study of the kinetics of its inhibition is difficult. The kinetic approach does appear to be warranted, however, because pepstatin, while a strong-binding inhibitor, is not an irreversible inactivator. This point was made clear in an experiment where pepstatin was added to a pepsin solution to achieve about 50% inhibition. After heat inactivation of the enzyme, the same amount of native pepsin was again added to this solution. This new mixture gave rise to a proteolytic activity which was equivalent to a 50% inhibition of the enzyme. It was concluded, therefore, that in the inhibition of pepsin, presumably also of other acid proteases, pepstatin neither forms a heat-stable covalent bond to the enzyme nor is it consumed in the process.

To further study the mode of inhibition of pepsin and other acid proteases by pepstatin, we tried to produce various pepstatin fragments and measure their kinetics of inhibition. These pepstatin fragments are produced by three methods: (a) complete acid hydrolysis yields free statine and other free amino acids; (b) partial acid hydrolysis yields a mixture of peptides; and (c) enzyme digestion with α-lytic protease yields the tetrapeptide, valyl-statyl-alanyl-statine. Purification by high-voltage paper electrophoresis from the above mixtures gives four statine-containing products in sufficient yield to allow further study: statine, alanyl-statine, valyl-statine, and valyl-statyl-alanyl-statine. The isolated products are acetylated and quantified, and the completeness of their acetylation is determined prior to the initiation of inhibition studies (10).

After plots of the data showed the inhibition to be competitive, the constants K_m, k_3, and K_i were calculated directly, as suggested by Cleland (11).

The results of kinetic experiments with N-acetylated statine and statine-dipeptides indicated that they are competitive inhibitors. This is shown in both the Dixon plot (12) and the Webb plot (13) for N-acetyl statine (Fig.1), using two substrates, globin and N-acetyl phenylalanyl-diodotyrosine. Similar competitive inhibition was observed for N-acetyl alanyl-statine (Fig.2) and N-acetyl valyl-statine. The K_i values calculated from these experiments are listed in Table I.

The nonacetylated statine and statyl dipeptides were also tested for inhibition. The K_i values are about ten times higher than the corresponding acetylated inhibitors (Table I). This is not surprising since it is known that the dipeptides with free NH_2-termini are poor substrates of pepsin as compared to the acetylated dipeptides.

The results described above have several interesting implications. The N-acetyl-statine and N-acetylated statyl dipeptides reversibly bind to the active center of pepsin. The binding of these inhibitors is considerably stronger than their amino acid and peptide analog, as the data in Table I clearly illustrates. While

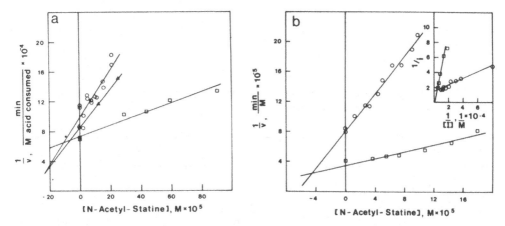

Figure 1. Kinetic plots of the inhibition of porcine pepsin by
N-acetyl-statine. a. Dixon plot (1/v versus [I]) for three
concentrations of globin, S_1 = 10.2 x 10^{-5}M, ($\square$); S_2 = 4.4 x
10^{-5}M, (Δ); S_3 = 3.1 x 10^{-5}M, (O). The rate of hydrolysis was
followed in a pH-stat at pH 2.1, 37°C. The enzyme concentration
was 1.5 x 10^{-6}M for S_1 and S_2 and 0.84 x 10^{-6}M for S_3. For the
purposes of this plot, the observed velocities for S_3 were converted to those expected if the enzyme concentration had been
1.5 x 10^{-6}M. b. Dixon plot for two concentrations of APDT,
S_1 = 10.9 x 10^{-5}M ($\square$), S_2 = 3.3 x 10^{-5}M (O). The pepsin concentration was 1 x 10^{-6}M. The rate of hydrolysis of APDT is
expressed as mmoles of diiodotyrosine produced per min per ml
of assay volume at pH 2.1, 37°C. b-inset. Webb plot (1/i
versus 1/I) of the same data where i = 1-(v_i/v_c) and v_c is the
velocity for the unhibited sample and v_i is the velocity for
an inhibited sample.

Table I

Kinetic parameters for inhibition of pepsin by statyl
derivatives and structurally similar peptides.

Inhibitor	Substrate	K_i
		M
N-acetylated		
N-Acetyl-statine	Globin	1.4×10^{-4}
	APDT	1.1×10^{-4}
N-Acetyl-alanyl-statine	APDT	5.6×10^{-6}
N-Acetyl-valyl-statine	Globin	6.8×10^{-6}
	APDT	2.8×10^{-6}
Non acetylated		
Statine	APDT	1.15×10^{-2}
Alanyl-statine		
Structurally similar inhibitors		
N-Acetyl-leucine		7×10^{-2}
N-Acetyl-phenylalanyl-phenylalanine		1.5×10^{-3}

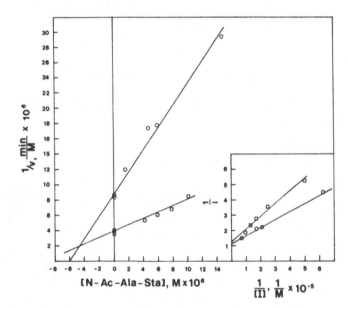

Figure 2. Kinetic plots of the inhibition of porcine pepsin by
N-acetyl-alanyl-statine. Dixon plot for two concentrations of
APDT, S_1 = 10.5 x 10^{-5}M, (■), S_2 = 3.0 x 10^{-5}M, (O). The concentrations of pepsin was 0.84 x 10^{-6}M.   Inset. Webb plot of
the same data.

N-acetyl-leucine is a structural analog of N-acetyl-statine (see Fig.3 for structure), its K_i (14) is about 600 x that for N-acetyl-statine. The same type of relationships are also observed for the K_i values of N-acetyl-valyl-statine (K_i = 2.8 x 10^{-6}) and N-acetyl-Phe-Phe (K_i = 1 x 10^{-3})(15); the nearest structural analog is N-acetyl-Val-Leu. No K_i value is available, however. This un-expected potency of these competitive inhibitors fits the criteria of the "transition state" inhibitors.

Another criterion of a "transition state" inhibitor is the structural similarity of the compounds to the transition state in the catalysis of the enzyme. This is illustrated in Figure 3, where the structural similarity of a statyl residue and one of the proposed transition states of peptic catalysis (16) can be readily seen.

These results suggest that statyl residues in pepstatin are the source of 'inhibitory potency'. The observation by Kunimoto et al. (4) that the O-acetyl pepstatin has a much reduced inhibitory acti-vity agrees with the current thesis.

Inhibition of N-acetyl-Val-Sta-Ala-Sta and Pepstatin

While the inhibition kinetics of smaller statyl compounds are fairly straightforward, the kinetics study of statyl-containing tetrapeptide, N-acetyl-Val-Sta-Ala-Sta, and pepstatin itself are much more complicated. First, it is more difficult to measure K_i of tight binding inhibitors. Moreover, if the statyl residues were the source of 'inhibitory potency', the presence of two statyl res-idues in each molecule of inhibitor casts further doubt on the the-oretical validity of kinetic parameters.

As shown in Figure 4, a plot of proteolytic activity of pepsin vs. the tetrapeptide concentration, according to Green and Work (17), produced a curve typical for a strong inhibitor, with a K_i of 2.5 x 10^{-9}M. The extrapolation of the activity lines to zero pro-teolytic activity, however, gave rise to pepsin/inhibitor ratios of about 2 to 1. Similar results are observed for pepstatin at pH 2.1 (10). These results seem to indicate that each of the two statyl residues in the tetrapeptide and pepstatin is capable of binding and inhibiting a pepsin molecule. For the tetrapeptide, a K_i value of 5 x 10^{-9}M was calculated on the basis of the concentration of statyl residues. This calculation is, of course, based on the assumption that the two statyl residues have equal binding strength and that the binding strength of a statyl residue is independent of the bind-ing strength of the second statyl residue in the molecule. These assumptions are experimentally difficult to verify at the present time. One thing is clear, however: the unusual pepsin/inhibitor.

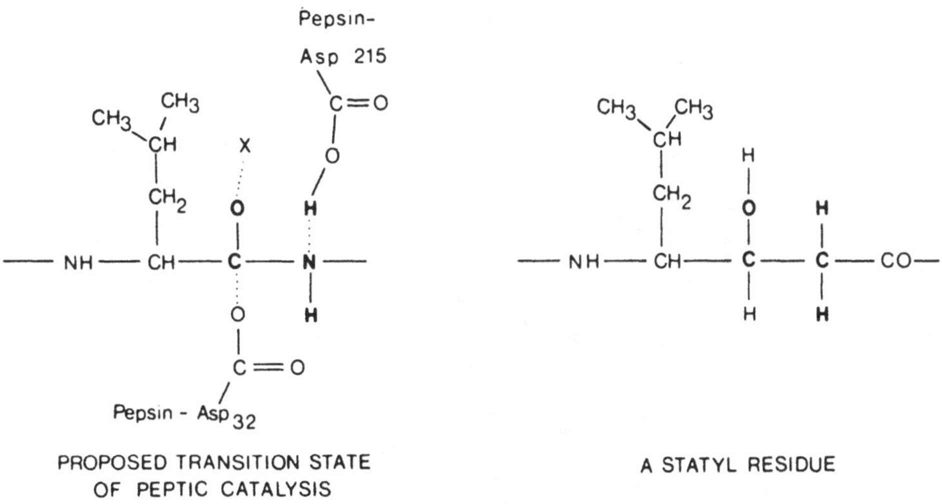

Figure 3. Schematic representation (left) of proposed transition state in peptic catalysis. Asp-32 and Asp-215 are respectively, the epoxide-end diazo-reactive aspartyl residues of the enzyme. X is a postulated electrophile. The carbonyl and amide groups of the substrate are shown in bold letters. The N-substitubed statine (right) is shown for comparison. The atoms in the statyl residue, which are equivalent to those shown in bold print in the proposed transition state, are also shown in bold letters.

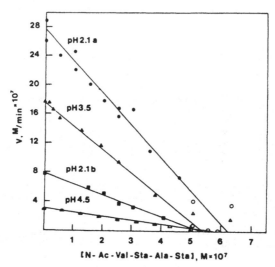

Figure 4.   N-Acetyl-Valyl-Statyl-Alanyl-Statine inhibition of
peptic activity at various pH values with APDT substrate.
Extrapolation to zero velocity gives the inhibitor concentration
for total inhibition. The pepsin concentration was 1 x 10^{-6}M.
Other experimental conditions and [E] / [I] ratios for total
inhibition for the various lines are: pH 2.1a, 1.1 x 10^{-4}M APDT,
[E]/[I] = 1.61; pH 2.1b, 0.29 x 10^{-4}M APDT, [E] / [I] = 1.89,
pH 3.5, 1.1 x 10^{-4}M APDT, [E]/[I] = 1.86; pH 4.5, 1.2 x 10^{-4}M
APDT, [E]/[I] = 1.77. Solid points were used to determine the
least squares lines.

ratio is not due to a slowness in equilibrium between enzyme and inhibitor, as suggested by Kunimoto et al. (4). We altered the mixing
order of enzyme, inhibitor, and substrate with virtually the same
inhibition kinetics.

On the other hand, there are good reasons to believe that the
basic modes of inhibition of the tetrapeptide and of pepstatin are
the same as that proposed above for the smaller statyl compounds
(Fig.3). The "competitiveness" and "transition state" hypotheses
for pepstatin are both supported by x-ray crystallographic results
(18 and Chapter 3 in this book) of R. chinensis in which the
reactive site of an active center-directed reagent, 1,2-epoxy-3-
(p-nitrophenoxy)propane, overlaps with the pepstatin site. Further
support comes from the studies of Rich et al. (19). Dideoxypepstatin,
which they synthesized, was shown to be a competitive inhibitor of
pepsin, with K_i = 2.1 x 10^{-7}M. This represents about a 2,000-fold
increase in K_i value caused by removal of the oxygen atoms from
3-position of statyl residues of pepstatin.

Inhibition of Other Acid Proteases

Since pepstatin inhibits all acid proteases, but not the proteases of other groups, a common transition state which is a result
of a common catalytic mechanism can obviously be suggested for this
group of proteases. We know from the work of Umerzawa et al.,
however, that the inhibitory potency of pepstatin differs for various
acid proteases (Ref.1, Table X). We confirmed this for pepstatin inhibition of a group of acid proteases (Table II). The inhibitory
potency of N-acetyl-Val-Sta-Ala-Sta also was found to vary widely.

An interesting question can thus be raised. If pepstatin is a
transition state inhibitor by virtue of the structure of statine, why
then is there a thousandfold difference in K_i values observed for
various acid proteases? Since there is no extensive kinetic data
available for the inhibition of a variety of acid proteases by
smaller statyl compounds, we can only speculate as to the answers.
It is unlikely that the catalytic transition state of various acid
proteases differs significantly. Not only are all these enzymes inhibited by pepstatin, they also are inhibited by the active center-
directed pepsin inactivators, such as diazoacetyl norleucine methyl
ester and 1,2-epoxy-(3-p-nitro-phenyoxy)propane. Although renin and
pepsin have been reported to be noncompetitively inhibited by pepstatin (20), questions have been raised on the interpretation of these
data; this has been extensively discussed by Rich (19). The differences in inhibitory potency among the various proteases could be
generated from a slight difference of the orientation of catalytic
groups in the active center. More likely, the binding of secondary
residues in longer statyl peptides may affect the orientation of the

Table II

Inhibition of Acid Proteases by Pepstatin and N-Acetyl-Tetrapeptide

Enzyme	% Inhibition[a] by Pepstatin	% Inhibition by N-Acetyl-Val-Sta-Ala-Sta
Porcine Pepsin	100 (1.2)[b]	72 (1.4)
Human Pepsin	100 (1.0)	65 (1.5)
Human Gastricsin	9 (1.3)	2 (1.6)
Acid protease from R. chinensis	10 (1.6)	0 (1.6)
Bovine Chymosin	46 (1.6)	N.D.[c]

[a]The % inhibition was determined using Assay I. The pepstatin concentration was 5×10^{-8}M and the N-Acetyl-Valyl-statyl-alanyl-statine concentration was 1.2×10^{-7}M.

[b]The form of the data is % inhibition (Enzyme concentration x 10^7M).

[c]N. D. means not determined.

statyl residue and thus its topographic relationship to the catalytic group in the active center.

ACKNOWLEDGMENT

This work was supported by Research Grants AM-01107 and AM-06487 from the National Institute of Health.

SUMMARY

Pepstatin is a low molecular weight, potent inhibitor specific for acid proteases with a K_i value of about 10^{-10}M for pepsin. The chemical structure of pepstatin is essentially a hexapeptide which contains two residues of an unusual amino acid, 4-amino-3-hydroxy-6-methylheptanoic acid (statine). The complete structure of pepstatin is isovaleryl-L-valyl-L-valyl-statyl-L-alanyl-statine.

To study its mode of inhibition, we prepared several derivatives and measured their kinetics of inhibition. Both N-acetyl-statine and N-acetyl-alanyl-statine are competitive inhibitors for pepsin with K_i values of 1.2×10^{-4}M and 5.65×10^{-6}M, respectively. The K_i value for N-acetyl-valyl-statine is 4.8×10^{-6}M. These statyl derivatives, therefore, are very strong inhibitors. The K_i value for N-acetyl-statine is 600-fold smaller than that of its structural analog N-acetyl-leucine. The derivative which contains two statyl residues in a tetrapeptide exhibits inhibitory properties which approach those of pepstatin itself. Other acid proteases, human pepsin, human gastricsin, renin, cathepsin D, the acid protease from R. chinensis and bovine chymosin, also are inhibited by pepstatin and its derivatives. We suggest that the statyl residue is responsible for the unusual inhibitory capability of pepstatin and that statine is an analog of the previously proposed transition state for catalysis by pepsin and other acid proteases.

REFERENCES

1. Umezawa, H. (1972) Enzyme Inhibitors of Microbial Origin, pp. 31, University Park Press, Baltimore
2. Kunimoto, S., Aoyagi, T., Nishizawa, R., Komai, T., Takeuchi, T., and Umezawa, H. (1974) J. Antibiotics 27, 413-418
3. Morishima, H., Takita, T., Aoyagi, T., Takeuchi, T., and Umezawa, H. (1970) J. Antibiotics 23, 263-265
4. Kunimoto, S., Aoyagi, T., Morishima, H., Takeuchi, T., and Umezawa. (1972) J. Antibiotics 25, 251-255
5. Aoyagi, T., Morishima, H., Nishizawa, R., Kunimoto, S., Takeuchi, T., Umezawa, H., and Ikezawa, H. (1972) J. Antibiotics 25, 689-694

6. Lundblad, R. L., and Stein, W. H. (1969) J. Biol. Chem. 244, 154-160

7. Bayliss, R. S., Knowles, J. R., and Wybrandt, G. B. (1969) Biochem. J. 113, 377-386

8. Tang, J. (1971) J. Biol. Chem. 246, 4510-4517

9. Chen, C. S., and Tang, J. (1972) J. Biol. Chem. 247, 2566-2574

10. Marciniszyn, J., Jr., Hartsuck, J. A., and Tang, J. (1976) J. Biol. Chem. 251, 7088-7094

11. Cleland, W. W. (1967) Advances in Enzymology 29, 1-33

12. Dixon, M. (1953) Biochem. J. 55, 170-171

13. Webb, J. L. (1963) Enzyme and Metabolic Inhibitors, Vol. I, pp. 58 Academic Press, New York

14. Inouye, K., and Fruton, J. S. (1968) Biochem. 7, 1611-1615

15. Knowles, J. R., Sharp, H., and Greenwell, P. (1969) Biochem. J. 113, 343-351

16. Hartsuck, J. A., and Tang, J. (1972) J. Biol. Chem. 247, 2575-2580

17. Green, N. M., and Work, E. (1953) Biochem. J. 54, 347-352

18. Subramanian, E., Swan, I. D. A., and Davies, D. R. (1976) Biochem. Biophys. Res. Commun. 68, 875-880

19. Rich, D. H., Sun, E., and Sengh, J. (1977) Biochem. Biophys. Res. Commun. 74, 762-767

20. McKown, M. M., Workman, R. J., and Gregerman, R. I. (1974) J. Biol. Chem. 249, 7770-7774

CHEMICAL MODIFICATION OF A PEPSIN INHIBITOR FROM THE ACTIVATION PEPTIDES OF PEPSINOGEN

P. M. Harish Kumar, Peter H. Ward, and Beatrice Kassell

From the Department of Biochemistry
Medical College of Wisconsin
Milwaukee, Wisconsin 53233, U.S.A.

A pepsin inhibiting substance formed during the activation of porcine pepsinogen has been known for many years (1). Based on the finding that in solutions more acid than pH 4, the loss of pepsinogen was not immediately accompanied by an equivalent increase in pepsin. Herriott postulated the scheme shown in Figure 1 for the autocatalytic activation of pepsinogen. The pepsin inhibitor was separated and partially characterized by Herriott (2) and by Van Vunakis and Herriott (3,4). At that time, the techniques of peptide purification and amino acid analysis had not been perfected, and their analysis does not agree with any part of the amino-terminal sequence later elucidated by Ong and Perlmann (5). In 1973, however, Anderson and Harthill (6) prepared an inhibitor almost corresponding in composition to the amino-terminal 16 amino acids of porcine pepsinogen. A homologous 17-amino acid peptide from bovine pepsinogen was reported jointly by Foltmann's group and by Kay and Kassell (7), while Kassell et al. (8) found a peptide from canine pepsinogen of similar composition to the first 14 amino acids of bovine pepsinogen. The high degree of homology of the amino-terminal portions of the pepsinogens of the three species, shown in Figure 2, makes it likely that the activation peptides are not simply "throw aways", but have a physiological role. We have therefore returned to the porcine activation peptides for further investigation, since porcine pepsinogen is readily available.

To study the basis for inhibiting activity, we have modified peptide 1-16 by guanidination of the lysine residues, carried out Edman degradation of the guanidinated inhibitor, and studied the activity of the resulting peptides.

211

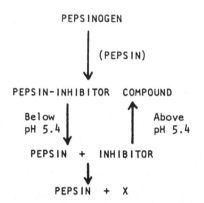

Figure 1. The scheme proposed by Herriott (1) to explain the delay in the formation of active pepsin upon activation of pepsinogen.

	1			5				10				15		

Porcine   LEU-VAL-LYS-VAL-PRO-LEU-VAL-ARG-LYS-LYS-SER-LEU-ARG-GLN-ASN-LEU

Bovine   SER-VAL-VAL-LYS-ILE-PRO-LEU-VAL-LYS-LYS-LYS-SER-LEU-ARG-GLN-ASN-LEU

Canine   ALA-ILE,VAL,LYS,VAL,PRO,LEU,VAL,ARG,LYS,LYS,SER,LEU,ARG

Figure 2. Homology in the structure of the amino-terminal portions of porcine, bovine, and canine pepsinogens.

EXPERIMENTAL PROCEDURE

Materials

Crystalline pepsinogen was purchased from Sigma. The sulfate of 2-methyl isourea was an Eastman product. Phenyl isothiocyanate and trifluoroacetic acid were Sequenal grade, 1 ml vials from Pierce Chemical Company. Solvents for Edman degradation were purified and checked for the absence of aldehydes according to Edman and Begg (9). Other chemicals were reagent grade.

Preparation of Peptide 1-16

The pepsinogen was first purified by chromatography on polyly-sine-Sepharose 4B at pH 6.5 (10). About 600 mg of purified pepsino-gen was activated at 0° and pH 2 for 90 sec at a protein concentra-tion of 6 mg/ml. The pH was raised to 3.5, 9.3 mg of pepstatin was added and the activation peptides were separated from the pepsin by the same method used by Dr. Kay in this laboratory for the bovine activation peptides (11,7), i.e. a polylysine-Sepharose 4B column equilibrated with ammonium formate buffer, pH 3.5. The mixture of peptides not retained by the column was lyophilized. The peptides were separated by chromatography on CM-Sephadex C-25 (Pharmacia) at pH 6.0 by a method similar to that of Anderson and Harthill (6), except that a volatile buffer was used (see the legend to Fig. 3 for details).

Guanidination of Peptide 1-16

A small-scale modification of the method of Chervenka and Wilcox (12) was used (cf.13). The free base form of 2-methyl-isourea, prepared by precipitating the sulfate with barium hydroxide, was added in 100-fold molar excess to 3 μmol of peptide 1-16 at 5^u and pH 10.3. After 72 hr at 5°, reagents were removed on a column of Sephadex G-15 (1 x 97 cm, equilibrated with 10 mM ammonium bicar-bonate). The peptide peak was lyophilized.

Amino acid analysis-Peptides and proteins were hydrolyzed in sealed evacuated tubes with redistilled constant boiling HCl at 110° for 24 and 48 hr,or with a 4 M methane sulfonic acid containing trypa-mine (Pierce) at 115° for 24 hr (14). Analyses were run on a modi-fied Beckman 120B amino acid analyzer (15).

Assay of pepsin-inhibiting activity-The activity of the pepsin used was calculated on the basis of 81% of active pepsin by compari-son of its hemoglobin-splitting activity with our purest pepsin preparation (10).

To determine inhibition of the milk-clotting activity of pepsin by peptide 1-16, the turbidimetric method of McPhie (16) was used with minor modification. We found that a more consistent standard curve for pepsin was obtained when the fresh skim milk purchased locally was first centrifuged at 12,000 x g for 10 min in the cold; 2 ml was used instead of 1 ml, diluted as described by McPhie with 44 ml of 0.2 M sodium acetate buffer, pH 5.3, and 4 ml of 0.1 M $CaCl_2$. The dilute milk was prepared each day, stored in the refrigerator, and a small amount equilibrated at 25° before each assay.

We confirmed the observation of Dunn et al.(17) that peptide 1-16 reacts with pepsin much more rapidly and completely at 37° than at 25°. The turbidimetric assay for pepsin, however, was less consistent at 37° than at the 25° temperature recommended by McPhie. We therefore first incubated the peptide with about 300 ng of pepsin in 1 ml of the pH 5.3 buffer for 15 min at 37°, then for 5 min at 25° before adding 2 ml of the diluted milk solution at 25° to start the assay. The standard curve for 50-300 ng pepsin was prepared in exactly the same way and rechecked with each new carton of milk.

Edman degradation. A sample of about 1 μmole was degraded essentially by the method of Peterson et al. (18), but with the entire procedure carried out inside the nitrogen chamber described by Meagher (19). At each step, following ether extraction of the amino acid thiazolinones (which were discarded), the residual peptide was dissolved in a few tenths ml of water. Samples were taken for subtractive amino acid analysis (20) and for determination of inhibiting activity. The solution was then evaporated to dryness under N_2 for the next step.

RESULTS

Results of the separation of the activation peptides, obtained as described above, are shown in Figure 3. Amino acid analysis of the peaks showed that the pepstatin was recovered in peak 1 and that peak 4 contained mainly peptide 1-16 (Table I). In addition to the amino acids listed, there were small amounts of histidine, glycine, and alanine. From the quantities of these amino acids and the known sequence of the total activation fragments (5,21,22), it was ascertained that the product was about 90% pure, the slightly high values for several amino acids being caused by contaimination with at least three other peptides. No further purification was considered considered necessary.

Analysis of the guanidinated inhibitor (Table I) indicated better than 96% conversion of the lysine residues to homoarginine, with no change in other amino acids.

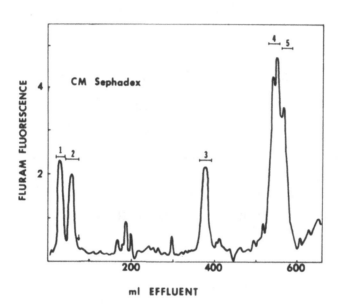

Figure 3. Separation of the peptides formed by activation of 600 mg of porcine pepsinogen on CM-Sephadex C-25. Column 1.2 x 42 cm. Starting buffer, 60 mM ammonium acetate, pH 6.0. Gradient applied at the arrow: 275 ml of starting buffer in a constant volume flask plus 0.8 M ammonium acetate in the reservoir. Fractions 2 ml. Peak 4 yielded peptide 1-16.

Table I

Amino Acid Composition of Peptide 1-16
and Its Guanidinated Derivative

Amino Acid	Residues per mole		
	Native	Guanidinated	Expected[a]
Lys[b]	3.42	0.13	3
Homoarg		3.25	
Arg	2.00	2.00	2
Asp	1.19	1.18	1
Ser	.90	1.02	1
Glu	1.14	1.10	1
Pro	1.27	1.20	1
Val	3.07	3.00	3
Leu[b]	4.26	4.17	4

[a]From the sequence of the amino terminal portion of porcine pepsinogen (Ref. 5).

[b]Slightly high values are due to about 10% contamination with
other activation peptides.

Table II

Edman Degradation of Guanidinated Peptide 1-16
Sequence: Leu-Val-Homoarg-Val-

Edman step	Ratio: Arg = 2		
	Leu	Val	Homoarg
0	4.17	3.00	3.25
1	3.24	3.01	3.18
2		2.05	3.44
3			2.55[a]
4		1.42	

[a]Incomplete extraction and reconversion during hydrolysis of
homoarginine thiazolinone.

Inhibition studies with the native inhibitor and its guanidinated derivative (Fig.4) demonstrated that the guanidinated inhibitor is a stronger pepsin inhibitor. For 50% inhibition, the molar ratio of peptide to pepsin was decreased by the modification from 9 to about 2.

The guanidinated inhibitor retained only one free amino group, which was on the terminal leucine. The absence of other reactive groups made it possible to subject this derivative to the Edman degradation. This in turn made it possible to determine the role of the amino acids near the amino terminus in the inhibiting activity. Four steps of Edman degradation were carried out as described in EXPERIMENTAL PROCEDURE. Another portion of the material was used as a control of nonspecific loss of activity due to exposure to the Edman reagents; this sample was subjected to identical treatment except for omission of the phenylisothiocyanate.

Table II shows the sequential removal of the first four amino acids from the guanidinated inhibitor. At step 3, the loss of the homoarginine residue did not appear to be complete. It is likely, however, that cleavage of the residue went farther than the results indicate, since the thiazolinone of arginine is known to resist extraction and homoarginine thiazolinone would be expected to behave similarly. Acid hydrolysis partially reconverts the thiazolinones or thiohydantoins to the free amino acids (23), accounting for the high value for homoarginine. Cleavage of the second valine in the fourth step proceeded farther than the apparent cleavage of the homoarginine, confirming this explanation. In the control for the Edman reaction (Table III) there was no significant change in amino acid composition.

The results of the inhibition assays of the degraded guanidinated inhibitor and its control are presented in Table IV. Removal of the amino terminal residues, with the exception of leucine, clearly had an adverse effect on the inhibition, but did not destroy it entirely. The fourth step was not tested, since degradation at the third step was probably incomplete.

DISCUSSION

We have shown in this paper that modification of peptide 1-16 of porcine pepsinogen by conversion of its three lysine residues to homoarginine increases its pepsin-inhibiting activity. The availability of an active guanidinated peptide made it possible to determine the role of the residues at the amino terminus. By step-wise Edman degradation, not possible with the native inhibitor containing other free amino groups, it was shown that the terminal leucine was not needed, but the neighboring residues were required for full activity.

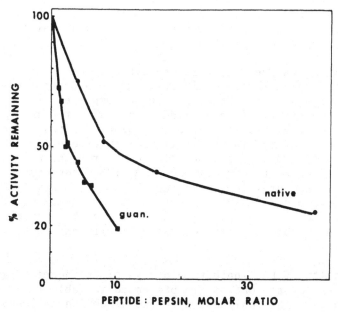

Figure 4.   Inhibition of 300 ng of pepsin by varying amounts
of native peptide 1-16 and its guanidinated derivative (guan.).

Table III

Guanidinated Peptide 1-16
Control for Edman Degradation
(No Phenylisothiocyanate)

Edman step	Ratio: Arg = 2		
	Leu	Val	Homoarg
0	4.17	3.00	3.25
1	3.99	3.05	3.22
2	3.98	3.03	3.30
3	3.76	2.90	2.99

Table IV

Guanidinated Peptide 1-16

Molar Ratio of Peptide to Pepsin
for 50% Inhibition

Edman Degradation			Control Ratio
Step	Amino Acid Removed	Ratio	
0	-	2	2
1	Leu	8	6
2	Val	80	7
3	Homarg	120	5

Dunn and his colleagues (17) have synthesized peptide 1-16 and 1-13; both were active inhibitors. Thus, the three carboxyl-terminal amino acids of peptide 1-16 are not essential.

The physiological role of activation peptides is of interest. The activation peptide of bovine trypsinogen, Val-Asp$_4$-Lys, inhibits the secretion of gastric juice (24). Thus, when food enters the intestine and gastric secretion is no longer needed, activation of trypsinogen provides an inhibitory peptide. To study the physiological role of the pepsin inhibitors, experiments would require peptides that are stronger inhibitors than the native peptides, as well as peptide inhibitors able to resist digestion by pepsin at acid pH. The guanidinated peptide is an approach to solving this problem.

ACKNOWLEDGMENT

This work was supported by U.S. Public Health Service Research Grant AM-09826 from the National Institute of Arthritis, Metabolism, and Digestive Diseases, and by Grant BMS-22102 from the National Science Foundation. We thank Dr. John Kay for the gift of a sample of pepsinogen and are grateful to Dr. William F. Minor of Bristol Laboratories for a gift of pepstatin.

SUMMARY

The peptide comprising the first 16 amino acids of porcine pepsinogen, prepared from the activation mixture, has been modified by guanidination of its three lysine residues to form homoarginine residues. The modified peptide is a better pepsin inhibitor than is the native peptide; for 50% inhibition of the milk-clotting action of pepsin at pH 5.3, the molar ratio of peptide to pepsin required is 9 for the native inhibitor and only 2 for the guanidinated inhibitor.

Stepwise removal by Edman degradation of the amino-terminal Leu-Val-Homoarg residues from the guanidinated inhibitor decreased the activity slightly at the first step and markedly at the second and third steps. Thus, all of the amino-terminal sequence except the leucine residue is necessary for full activity.

REFERENCES

1. Herriott, R. M. (1938) J. Gen. Physiol. 22, 65-78
2. Herriott, R. M. (1941) J. Gen. Physiol. 24, 325-338
3. Van Vunakis, H., and Herriott, R. M. (1956) Biochim. Biophys. Acta 22, 537-543

4. Van Vunakis, H., and Herriott, R. M. (1957) Biochim. Biophys.
    Acta 23, 600-608
5. Ong, E. B., and Perlmann, G. E. (1968) J. Biol. Chem. 243,
    6104-6109
6. Anderson, W., and Harthill, J. E. (1973) Nature 243, 417-419
7. Harboe, M., Andersen, P. M., Foltmann, B., Kay, J., and
    Kassell, B. (1974) J. Biol. Chem. 249, 4487-4494
8. Kassell, B., Wright, C. L., and Ward, P. H. (1976) Protides of
    the Biological Fluids - 23rd Colloquium (Peeters, H. ed)
    pp. 541-544, Pergamon Press, Oxford and New York
9. Edman, P., and Begg, G. (1967) Eur. J. Biochem. 1, 80-91
10. Nevaldine, B., and Kassell, B. (1971) Biochim. Biophys. Acta
    250, 207
11. Kay, J. (1972) Fed. Eur. Biochem. Soc. Proc. Meet. Abstr. 458
12. Chervenka, C. H., and Wilcox, P. E. (1956) J. Biol. Chem. 222,
    635-647
13. Kassell, B., and Chow, R. B. (1966) Biochemistry 5, 3449-3453
14. Simpson, R. J., Neuberger, M. R., and Liu, T. Y. (1976) J. Biol.
    Chem. 251, 1936-1940
15. Eick, H. E., Ward, P. H., and Kassell, B. (1974) Anal. Biochem.
    59, 482-488
16. McPhie, P. (1976) Anal. Biochem. 73, 258-261
17. Dunn, B. M., Moesching, W. G., Trach, M. L., Nolan, R. J., and
    Gilberts, W. A. (1976) Fed. Proc. 35, 1463
18. Peterson, J. D., Nehrlich, S., Oyer, P. E., and Steiner, D. F.
    (1976) J. Biol. Chem. 247, 4866-4871
19. Meagher, R. B. (1976) Anal. Biochem. 67, 404-412
20. Konisberg, W., and Hill, R. J. (1962) J. Biol. Chem. 237,
    2547-2561
21. Pedersen, V. B., and Foltmann, B. (1976) FEBS Lett. 35, 255-256
22. Stepanov, V. M., Baratova, L. A., Pugacheva, I. B., Belyanova,
    L. P., Revina, L. P., and Timokhina, E. A. (1976) Biochem.
    Biophys. Res. Commun. 54, 1164-1170
23. Fraenkel-Conrat, H., Harriss, J. I., and Levy, A. L. (1955) in
    Methods of Biochemical Analysis, Vol. II (Glick, D. ed)
    pp. 359-425, Interscience, New York
24. Abita, J. P., Moulin, A., Lazdunski, M., Hage, G., Palasciano,
    G., Brasca, A., and Tiscornia, O. (1973) FEBS Lett. 34, 251-255

NOTE ADDED IN PROOF

A sample of peptide 1-16 was tested for inhibition of human
kidney renin by Dr. B. Leckie of the MRC Blood Pressure Unit,
Western Infirmary, Glasgow, Scotland.

At a renin concentration of 4.1 μunits per ml and a peptide
concentration of 2.29 μM, there was no inhibition of angiotensin I
released from ox renin substrate. Similarly, there was no inhibition
of the renin enzymatic activity in male human plasma at the same
inhibitor concentration.

ACID PROTEASES IN VARIOUS BIOLOGICAL SYSTEMS

RENIN AND PRECURSORS: PURIFICATION, CHARACTERIZATION, AND STUDIES ON ACTIVE SITE

Tadashi Inagami, Kazuo Murakami, Kunio Misono, Robert J. Workman, Stanley Cohen and Yasunobu Suketa

From the Departments of Biochemistry and Medicine
Vanderbilt University School of Medicine
Nashville, Tennessee 37232

Renin is an enzyme elaborated in juxta glomerular cells of the kidney and released into the blood stream by various stimuli. Although it is an endopeptidase, its function is strictly limited to the formation of angiotensin I from angiotensinogen by the cleavage of a unique leucyl-leucine peptide bond in this substrate molecule (Fig.1). Angiotensin I is converted into the octapeptide angiotensin II by a carboxydipeptidase, known as converting enzyme, and thenceforth to the heptapeptide angiotensin III by an aminopeptidase. Not only is angiotensin II the most potent pressor substance known, but both angiotensin II and III also stimulate the adrenal cortex to release the mineral corticoid aldosterone. Thus, renin triggers a chain of events aimed at elevating the blood pressure. Because of its strategically important position in the renin-angiosin-aldosterone system, the activity of renin in the circulation is tightly regulated by intricate multiple feedback control mechanisms. Excellent reviews on the control of renin release have been published recently (1-4).

Past studies of renin were severely hampered by the lack of a pure preparation of this important enzyme. Attempts at purification were hindered by the extremely low concentration of renin in the kidney and the increasing instability of the enzyme as it attained greater degrees of purity.

MOUSE SUBMAXILLARY GLAND RENIN

It was fortunate that renin purification studies by our group began with an organ which contains extraordinary high concentration of renin. The submaxillary gland of the male adult mouse (5) and

225

FORMATION OF ANGIOTENSINS

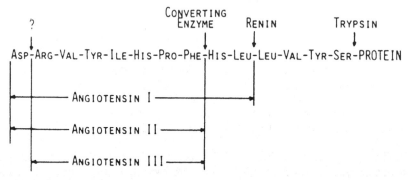

Figure 1. Role of renin in the formation of angiotensins.

Table I
Molecular properties of mouse submaxillary
gland renin A

Molecular weight	
Sedimentation equilibrium, low speed	37,100
high speed	35,900
Gel filtration	43,000
Isoelectric point	5.4
$E^{1\%}_{1cm, 280nm}$	10.0
Carbohydrate	not detectable

rat (6) had been known to contain a large amount of renin-like enzyme. Starting with a crude extract of mouse submaxillary gland, 5 column chromatography procedures afforded electrophoretically homogeneous preparations of the submaxillary gland enzyme (7). After the final purification step on a carboxymethyl cellulose column, five peaks were obtained (Fig.2). All the isoenzymes were closely related in their immunological, molecular, and enzymic properties with slightly different isoelectric points. The main component designated as "mouse submaxillary gland renin A" was crystallized for crystallographic studies currently being conducted in collaboration with Dr. G. Lenhert. Properties of this enzyme are summarized in Table I. Amino acid composition as shown in Table II is very similar to that of porcine kidney renin. One conspicuous difference between porcine renin and the mouse submaxillary gland enzyme is the lack of carbohydrate in the submaxillary enzyme, whereas the renal enzymes are glycoproteins containing glucosamine and possessing specific binding affinity to concanavalin A.

PURIFICATION OF RENAL RENIN

As we mentioned, the purification of renal renin had presented insurmountable difficulties in the past (8-17). Recently, several special techniques have been devised and applied to this difficult problem, including affinity chromatography using synthetic inhibitor peptides of renin as affinity ligands (17,20). Burton, et al. synthesized specific renin inhibitors for this purpose (21). These synthetic peptides, however, did not possess sufficiently strong affinity for renin to sequester it from a crude kidney extract.

We were able to prepare a different type of affinity gel using a bacterial fermentation product, the acylated pentapeptide pepstatin (22). The choice of pepstatin as an affinity ligand for renin was based on the partial structural similarity between pepstatin and renin substrate, as will be discussed later, as well as on the reports (23-26) that pepstatin is inhibitory to renin. Corvol et al. (27) and we (28,29) prepared pepstatin-aminoalkyl-agarose gels (Fig.3). This affinity column was highly effective in selectively absorbing renin present in minute quantity in crude kidney extracts containing vast amounts of non-renin proteins.

There was another serious obstacle to the purification of renin. The reported instability of renin during its purification (12), particularly in its purer state, has been the major deterrent to the numerous attempts of renin purification in the past. We thought it likely that this instability was due to proteolytic destruction by contaminating proteases. Since elimination of these contaminants by physical separation methods was extremely difficult, they were selectively inactivated chemically by treating crude renal extracts with a cocktail of protease inactivators, diisopropyl-phosphoro-

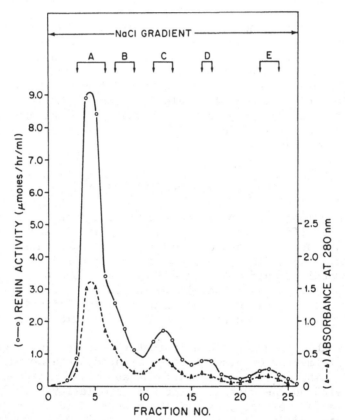

Figure 2. Chromatographic elution pattern of mouse submaxillary gland renin at its final purification step from
a carboxymethyl cellulose column at pH 5.4. Renin activity
(-0-) was determined by the method of Reinharz and Roth (62).
Protein concentration (--Δ--) was measured by absorbance at
280 nm. Ref. (54).

Table II
Amino acid compositions of mouse
submaxillary gland renin and porcine renal renin

Amino acids	Submax renin	Porcine renin	Amino acids	Submax renin	Porcine renin
	Residues/molecule			Residues/molecule	
Lys	12	11	Val	35	34
His	8	5	Met	7	4
Arg	9	9	Ile	15	12
Asp	27	26	Leu	33	30
Thr	27	26	Tyr	17	16
Ser	32	33	Phe	17	17
Glu	27	31	Trp	5	5
Pro	19	18	½Cys	4	2 ~ 4
Gly	35	34	GlcN*	0	3
Ala	14	16	GalN*	0	0

*GlcN and GalN are glucosamine and galactosamine respectively.

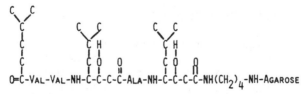

Figure 3. Structures of affinity gels used for renin purification.

fluoridate (0.05 mM), phenylmethanesulfonyl fluoride (2 mM), Na-
tetrathionate (5 mM), and EDTA (5 mM). Extraction and initial
fractionation were carried out by the modification of Rubin's method
(3). Minced, freeze-dried, pulverized, and ether-defatted powder
of porcine kidney cortex was extracted in 30% (v/v) methoxyethanol
containing the above-mentioned protease inactivators. Renin
activity in the extract was adsorbed to DEAE-cellulose at neutral
pH, eluted in 0.1 M Na-acetate containing 0.2 M NaCl at pH 4.8, and
precipitated with 72% saturated ammonium sulfate. The precipitate
was dissolved in 0.01 M pyrophosphate buffer, pH 6.5, containing
0.1 M NaCl and applied to a column of pepstatin-aminohexyl-agarose.
After exhaustive washing with the same buffer and with 0.1 M Na-
acetate, pH 5.2, containing 1 M NaCl, renin was eluted with 0.1 M
acetic acid, (pH 3.2). Fractions containing renin activity were
concentrated by pressure filtration and purified further by succes-
sive chromatography on columns of Sephadex G-75, DEAE-cellulose and
CM-cellulose. The final purification on the CM-cellulose column
produced a symmetrical main peak of renin activity superimposable on
a symmetrical protein concentration peak (Fig.4). Purity of this
preparation was established by a number of criteria such as disc
electrophoresis in the absence and presence of SDS, isoelectric
focusing and sedimentation equilibrium studies. The result of the
purification of porcine renin is summarized in Table III. A final
133,000-fold purification was achieved with a remarkably high overall
yield of 19%. The table illustrates yields beginning with 6 kg of
frozen porcine kidneys. In actual practice, three batches of product
from the affinity chromatography were pooled, and the combined
fractions were purified by the above-described procedures. This
method produced solutions of higher concentration. Since purified
renin tends to be more stable the higher its concentration, the pool-
ing procedure resulted in higher yields of final product. Although
there is no doubt that affinity chromatography on the pepstatin
agarose gel made the single most important contribution of a 300-fold
purification to these isolation studies, the product obtained by this
method was not homogeneous. Three additional column chromatography
steps were necessary to achieve an additional 30-fold purification.
This scheme resulted in the first complete purification of renin
since its discovery by Tigerstedt and Bergman in 1898.

PROPERTIES OF RENINS FROM PORCINE, RAT, AND HUMAN KIDNEYS

Contrary to previous reports, the porcine renin preparation
thus obtained was quite stable. When a solution containing 0.17 mg
of renin per ml in 0.02 M Na phosphate buffer, pH 6.3, containing
0.1 M NaCl, was stored at 4° or frozen at -20°, no detectable loss
of activity was observed over a period of eight weeks. It is likely
that the chemical inhibition of contaminating proteases and the
relatively high concentration of the enzyme protein contributed to
the stability.

Table III
Purification of Porcine renin

Purification step	Specific activity*	yield%
Frozen porcine kidneys		
Minced, freeze-dried, pulverized, defatted	0.002	100
Extraction in 30% methoxyethanol	0.009	93
Adsorption on DEAE-cellulose	-	-
Elution in 1 M NaCl at pH 4.8	-	-
Precipitation with $(NH_4)_2SO_4$ at 72% saturation	0.025	74
Affinity chromatography	9.3	55
Gel filtration	84.7	50
DEAE-cellulose chromatography	260	31
CM-cellulose chromatography	267	19

*μg angiotensin I formed /μg enzyme /h at pH 6.0

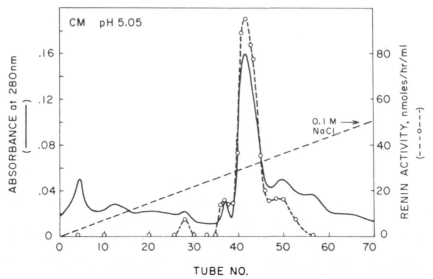

Figure 4. Chromatographic elution pattern of porcine renal renin at its final purification step from a carboxymethyl cellulose column at pH 5.05. Renin activity (--O--) was determined by the method of Reinharz and Roth (62), and the protein concentration (———) was determined by absorbance at 280 nm. Ref. (7).

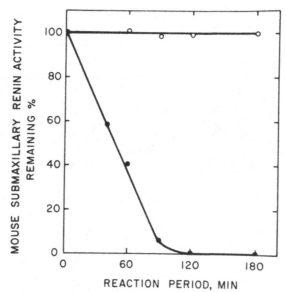

Figure 5. Inactivation of mouse submaxillary gland renin by diazo-acetyl-D,L-norleucine methyl ester (2 mM). The reaction was carried out in the presence (0) and absence (0) of 1 mM cupric acetate at pH 5.4 and 14°. Ref. (38).

The amino acid composition of porcine renin is closely analogous to that of mouse submaxillary gland renin (Table II). The porcine renal renin, however, contains glucosamine amounting to approximately 1.2% of the total weight. The molecular weight of this enzyme protein is estimated to be 42,500 by gel filtration on a column of Sephadex G-100 and 36,400 by sedimentation equilibrium.

The above purification method was applied with some modification to the isolation of rat renal renin to homogeneous states. Rat renin has a much weaker affinity for the pepstatin-aminohexyl-agarose gel than porcine renin (31). Human renin was also partially purified by a similar technique (32). Both rat and human renins are glycoproteins since they are bound by concanavalin A and released by O-methyl-D-glucose.

ACTIVE SITE OF RENIN

Although renin is a protease, its highly restrictive substrate specificity makes it of particular interest to enzymologists. Because of its uniqueness we were interested in identifying the functional groups essential for its catalytic function. Previous studies designed to investigate this problem used specific inactivators of proteases such as diisopropylphosphorofluoridate (33,34), p-hydroxy-mercuribenzoate (34), and EDTA (33), and showed that renin is refractory to these reagents, indicating that renin does not belong to the classes of serine, cysteine, or metallo-proteases. Our preliminary studies with group-specific reagents for functional groups of proteins such as methyl acetimidate, a specific reagent for primary amino groups, and diethyl pyrocarbonate, a specific reagent for amino and imidazole groups, were also negative, indicating that the amino and imidazole groups of renin are not essential for its enzymatic activity.

Aliphatic diazo compounds (35,36) and epoxides (37) have been known to react with carboxyl groups in the active site of acidic proteases. When diazoacetyl-D,L-norleucine methyl ester (2 mM) reacted with mouse submaxillary gland renin at pH 5.4 in the presence of 1 mM cupric acetate, a rapid and complete inactivation of the enzyme was observed as illustrated by the solid circles in Figure 5. The cupric ion was essential for this inactivation since the diazo compound alone in the absence of the cupric ion failed to inactivate the enzyme, as shown by the open circles in Figure 5 (38). Porcine renal renin was also inactivated rapidly in a similar manner as shown in Figure 6. Other diazoacetyl derivatives such as diazoacetyl-glycine methyl ester and diazoacetyl leucine methyl ester were also effective inactivators of renin. Mckown and Gregerman have been able to inactivate human renin by a similar technique using diazoacetyl-D,L-norleucine methyl ester (39).

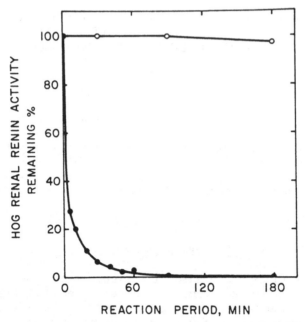

Figure 6. Inactivation of porcine renal renin by diazoacetyl-
D,L-norleucine methyl ester (4 mM), in the presence (●) and
absence (0) of 2 mM cupric acetate at pH 6.2 and 14°. Ref (38).

Table IV
Optimal pH of renin activity on different substrates

Renin	Substrate	Optimal pH
Mouse submaxillary gland renin	octapeptide*	5.4
	tetradecapeptide	6.5
	sheep plasma substrate	8.3
Rat renal renin	octapeptide*	5.5
	rat plasma substrate	5.5-6.5
	porcine plasma substrate	5.5-6.5
Porcine renal	octapeptide	5.5
	porcine plasma substrate	5.5-6.5

*Benzyloxycarbonyl-Pro-Phe-His-Leu-Leu-Val-Tyr-Ser-β-naphthylamide

The submaxillary gland enzyme which had been completely inactivated by Cu^{++}-diazoacetyl-D,L-norleucine methyl ester was found
to contain 1.2 moles of covalently bound norleucine per mole of
enzyme upon amino acid analysis. This result suggested that an
equimolar stoichiometric reaction between the catalytically
essential group of the enzyme and the diazo inhibitor took place.

In an attempt to identify this functional group, we treated the
modified enzyme with 0.2 M hydroxylamine at pH 9.0 for 24 hr at
room temperature in the presence of 8 M urea. Upon exhaustive
dialysis followed by acid hydrolysis, the enzyme protein thus treated
was found to be devoid of norleucine. This result indicated that the
norleucine-containing compound was bound to renin by an ester linkage
involving a carboxyl group of the enzyme. Thus we were able to show
that renin was inactivated by diazoacetyl-D,L-norleucine methyl
ester in the presence of cupric ion, resulting in stoichiometric
incorporation of glycolyl-D,L-norleucine methyl ester by the esterification of a catalytically essential carboxyl group of renin.

By analogy to similar reactions observed with various acidic
proteases (for example 41-43), it is highly likely that the reaction
between renin and aliphatic diazocompounds involve the β-carboxyl
group of an aspartyl residue. Furthermore, renin is inactivated by
epoxides such as 1,2-epoxy-3-phenoxy-propane (40), which is also an
inactivator of acidic proteases. These results, together with the
fact that the acidic protease inhibitor pepstatin is a potent inhibitor of renin, have led Inagami and Misono (38) and Mckown and Gregerman (39) to propose that renin can be classified as an acidic
protease.

As Gregerman has pointed out, however, the term "acidic protease"
applied to renin is misleading since this enzyme is normally functional over a range of pH between 5 and 8. Our studies have indicated
that the pH optimum depends upon the substrate. When 3 substrates
of different molecular size are reacted with mouse submaxillary
gland renin, a wide range of pH optima is observed. As summarized
in Table IV, the smallest substrate, the octapeptide of Roth and
Reinharz (44), is most rapidly hydrolyzed at pH 5.4, the tetradecapeptide (45) has a pH optimum near pH 6.5 and sheep plasma substrate
reacts most rapidly at a pH above 8. Similarly, pure rat renin has
a pH optimum near 3.5 for the octapeptide substrate, whereas rat
and porcine angiotensinogens react optimally at or near pH 6.0.
The variation of optimal pH over such a wide range depending on the
molecular size of the substrate suggests that the pH optimum is
determined by an interaction involving a relatively large surface
area of the enzyme and substrate molecules.

Since renin seems to share common properties of catalytic
function with other acidic proteases, the unique and highly restricted substrate specificity of renin may reside in a very selective

mechanism of enzyme substrate interaction of binding. Skeggs, et al.
(46) have examined various peptides containing part of the structure
of the tetradecapeptide substrate (45) for their ability to function
as substrates of renin. The octapeptide sequence Pro-Phe-His-Leu-Leu-
Val-Tyr-Ser was considered to constitute the minimal requirement
for a renin substrate.

We have examined the inhibitory activity of the peptides listed
in Table V in order to estimate their binding affinity, since the
inhibition constant K_i is identical to the dissociation constant of
the enzyme-inhibitor complex. The results listed indicate that the
binding affinity of the inhibitors is contributed mainly by the
short segment on the carboxyl-terminal side of the tetradecapeptide
Little, if any, affinity was exhibited by angiotensin I and II which
represent the amino-terminal segment of the substrate molecule.

Based on this information, we designed a series of active site-
directed inactivators of renin. Leucyl residues in the peptide
Leu-Leu-Val-Tyr-Ser were replaced with a halogenated structural
analog of L-leucine, α-bromo-L-isocaproic acid (BIC) (47). Among
the most efficient inhibitors synthesized was L-BIC-Val-Tyr-NH$_2$,
an analog of Leu-Val-Tyr-NH$_2$ as shown in Table VI. Although the
rate of inactivation was rather slow, complete inactivation resulted
with stoichiometric incorporation of the inhibitor molecule which
was bound to the enzyme through an esteric linkage involving a
carboxyl group of the enzyme, as indicated by the release of radio-
labeled inhibitor by hydroxylamine. These studies on the relationship between the structure of peptides and their affinity to renin
were useful in devising the affinity gels for renin discussed earlier.

HIGH-MOLECULAR-WEIGHT-FORMS OF RENIN

Recently a number of investigators have reported the presence
of multiple forms of renin in crude extracts of kidney and plasma
(48-62). In addition to the enzyme with a molecular weight of
40,000, renins with higher molecular weight have been observed.
These high-molecular-weight renins could be activated to a higher
level of activity by various treatments such as acidification,
exposure to low temperature or limited proteolysis by proteases.
Controversies exist concerning the mechanism of the activation.
For example, in certain instances the activation involved reduction
in molecular weight whereas in other instances there was no
appreciable molecular weight change. Since the activation process
of a renin precursor potentially can have a profound effect on the
plasma renin activity, we initiated studies on high-molecular-weight
forms of renin.

Table V

Interaction of peptides representing part of the structure of the
tetradecapeptide substrate with mouse submaxillary gland renin

Peptides	K_i
	mM
Asp-Arg-Val-Tyr-Ile-His-Pro-Phe-His-Leu-Leu-Val-Tyr-Ser	
Leu-Leu-Val-Tyr-OMe	0.2
Leu-Leu-Val-Phe-OEt	0.2
Leu-Leu-Val-OMe	5.7
Leu-Val-Tyr-OMe	4.4
Asp-Arg-Val-Tyr-Ile-His-Pro-Phe-His-Leu	No inhibition at 1.5 mM
Asp-Arg-Val-Tyr-Ile-His-Pro-Phe	No inhibition at 1 mM

Table VI

Active site-directed inactivators of renin

Inactivators	Rate constant of inactivation
	h^{-1}
L-BIC-Leu-Val-Tyr-Ser-OH	0.083
D-BIC-Leu-Val-Tyr-Ser-OH	0.015
L-BIC-Val-Tyr-Ser-OH	0.082
D-BIC-Val-Tyr-Ser-OH	0.032
L-BIC-Val-Tyr-NH_2	0.20
D-BIC-Val-Tyr-NH_2	0.009
L-BIC-Leu-OCH_3	0.095

*Determined at pH 5.6 using mouse submaxillary gland renin. The rate constant of inactivation is expressed as pseudo-first-order rate constant.

These studies began in an attempt to reconcile discrepancies in molecular weight estimates of renin in purified form and in crude renal extracts. When the freeze-dried, defatted powder from porcine kidney was extracted at neutral pH in 30% (v/v) methoxyethanol containing the mixture of protease inactivators described earlier, and the extract was passed through a calibrated column of Sephadex G-100, renin activity was distributed in two areas corresponding to a minor peak of molecular weight 140,000 and a major peak of 61,000. This observation differed from that made with the purified porcine renin preparation which eluted at fractions corresponding to a molecular weight of 42,500, suggesting that conversion of high-to low-molecular-weight forms of renin took place during the course of purification. Subsequently, this conversion occurred during the elution of renin from the affinity column by acidification to pH 3.2 with 0.1 M acetic acid.

Attempts were made to avoid acidification by devising a gel whose affinity to renin was attenuated compared to the original pepstatin-aminohexyl-agarose. This was accomplished by reducing the length of the diamino alkane spacer group used as the bridge between pepstatin and agarose, and diaminobutane was used to prepare the pepstatin-aminobutyl-agarose gel (Fig.3). Renin activity in the crude extract of porcine kidney could be adsorbed on to this gel at pH 5.5 and eluted with 2 M urea, whereas the previously used pepstatin-aminohexyl-agarose gel required 4 M urea, a condition which rapidly denatured renin activity. Renin eluted with 2 M urea consisted of 3 different molecular weight species when fractionated on a column of Sephadex G-150 (Fig.7). Fractions indicated as BB correspond to a molecular weight of 140,000; this species of renin are designated big-big renin. The middle fractions indicated as B have a molecular weight of 61,000 and are designated big renin. Fractions S represent the renin isolated previously whose molecular weight is 36,400 by sedimentation-equilibrium and 42,500 by gel filtration. Apparently, conversion of the high-molecular-weight forms to this low-molecular-weight form could not be prevented completely even under the mild conditions employed.

The big renin and the big-big renin were purified by the method shown in the following scheme:

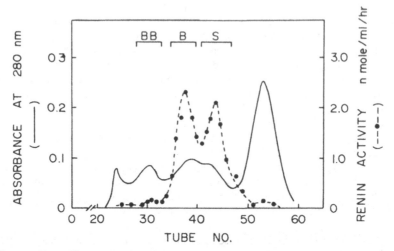

Figure 7. Chromatographic elution pattern of porcine big-
big renin (BB), big renin (B) and small renin (S) from a
Sephadex G-150 column. The renin preparation obtained
after affinity chromatography on a pepstatin-aminobutyl-
agarose gel was the starting material. The renin activity
(-●-) and protein concentration (——) were determined as
in Figure 2 and Ref. (60).

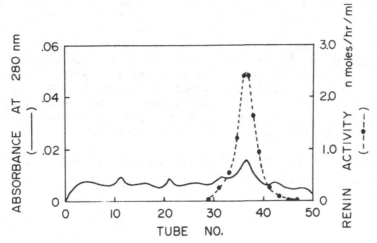

Figure 8. Chromatographic elution pattern of the final
stage of purification of porcine kidney big renin on a
DEAE-cellulose column at pH 6.1. Renin activity (-●-)
and protein concentration (——) were determined as in
Figure 2. Ref. (60).

```
            Kidney powder
               ↓
            Extract in 30% acetone with protease inactivators
               ↓
            Affinity chromatography on pepstatin-aminobutyl-agarose
               ↓
            Sephadex G-150
               ↓
  ┌────────────────────────────────────────────────────┐
  ↓                                                      ↓
CM-cellulose,pH 5.5                    DEAE-cellulose,pH 6.1
  ↓                                                      ↓
DEAE-cellulose,pH 6.1                  DEAE-cellulose,pH 6.5

Big renin                                  Big-big renin
```

The big renin thus isolated (Fig.8) was homogeneous by polyacrylamide gel electrophoresis (Fig.9B). The last step of the chromatographic purification separated the big-big renin into two components
which were designated as big-big renin I and II (Fig.10, 1st peak
and 2nd peak, respectively). Each was homogeneous as examined by
polyacrylamide gel electrophoresis (Fig.9)BB-I and BB-II. Re-
examination of the molecular weights of purified big and big-big
renin by a calibrated column of Sephadex G-150 confirmed that the
molecular weights estimated in the beginning of the isolation work
were preserved. The yield of the products was low. Twelve kilo-
grams of porcine kidney produced 0.16 mg of big renin and 0.25 mg
of big-big renin II.

    Properties of these high-molecular-weight renins are summarized
in Table VII. The fact that the pI of the big-big renin is clearly
different from that of the other species eliminates the possibility
that big-big renin is an oligomeric complex of the smaller species.
The inverse relationship between the molecular weights and the
specific activities of these renins strongly suggests the existence
of an enzyme-precursor relationship as shown below:

big-big renin ──────────────→ big renin ──────────────→ small renin

    In the context of this scheme, big-big renin I with a higher
specific activity than big-big renin II may very well be another
intermediate product of activation. Although big-big renin II is
only 1/2,000 as active as the small renin, it is possible that a
completely inactive renin precursor exists.

    Very preliminary studies using SDS-polyacrylamide gel electro-
phoresis have indicated the presence of a 42,000 molecular weight

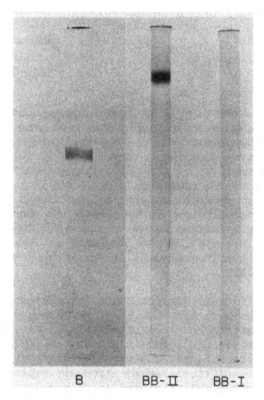

Figure 9. Polyacrylamide gel electrophoresis of purified
porcine big renin (B), big-big renin I (BB-I) and big-big
renin II (BB-II). Ref. (60).

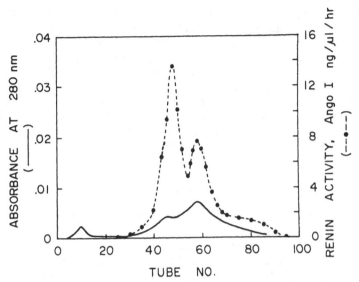

Figure 10. Chromatographic elution pattern of the final stage of purification of porcine kidney <u>big-big</u> renins on DEAE-cellulose at pH 6.5. Renin activity (-●-) and protein concentration (——) were determined as in Figure 2. Ref. (60).

Table VII

Properties of high-molecular-weight forms of porcine renin

	molecular weight**	specific activity**	isoelectric point
<u>Small</u> renin	42,000	267	5.2
<u>Big</u> renin	61,000	55	5.6
<u>Big</u>-<u>big</u> renin I	140,000	.5	7.3
<u>Big</u>-<u>big</u> renin II	140,000	.13	7.3

*Determined by gel-filtration on a calibrated column (1.5 x 100 cm) of Sephadex G-150.
**Rate of angiotensin production at pH 6.0 and 37° in µg angiotensin I formed/h/µg enzyme.

renin in both the big and big-big renin, probably as a non-covalently
bound component of these high-molecular-weight species of renin.
This observation agrees with those of Boyd (51) and of Leckie and
McConnell (55). It was not possible, however, to activate pure big
renin under the conditions which have been shown to activate big
renin in crude systems (48,50,51,62). The observation of big-big
renin in our study was the first discovery of renin of this molecular
size. The existence of a renin with a very low specific activity,
much lower than that of big renin, suggests the possible presence
of a totally inactive, high-molecular-weight zymogen of renin. This
big-big renin, however, seems to be different from the very-high-
molecular-weight substance observed by Barrett et al. (58).

SUMMARY

For the first time since its discovery in 1898 by Tigerstedt
and Bergman, renin from various sources has been purified by homogeneity, enabling the performance of biochemical studies with pure
enzyme which were previously impossible. Mouse submaxillary gland
renin was purified by conventional methods. Porcine renal renin was
purified by the combination of affinity chromatography on pepstatin-
aminohexyl-agarose gel and conventional chromatographic procedures.
Rat renin also was purified by an analogous method. Human renin was
partially purified.

Molecular properties of the purified renins were characterized.
Porcine renal renin and mouse submaxillary gland renin were found
to have similar amino acid compositions, molecular weights (approximately 37,000) and isoelectric points (5.2-5.7). The major difference between these renins is that the renal renins are glycoproteins
containing glucosamine whereas the mouse submaxillary gland renin
does not contain detectable carbohydrate.

Enzymological studies using various group-specific inactivators
of proteases indicated that renin belongs to the class of acidic
proteases. The structural requirements of renin substrates were
studied to elucidate the exceptionally selective substrate specificity of this peptidase whose only known function is to generate
angiotensin I. These studies have led to the design and synthesis
of active site-directed inhibitors of renin.

Studies of the high-molecular-weight forms of renin have been
conducted. We have discovered that renin in the kidney is all in
high-molecular-weight forms and that it does not exist in its most
active low-molecular-weight forms in this organ. Furthermore, there
are two high-molecular-weight species, big renin of 61,000 and big-
big renin of 140,000 molecular weight.

A special affinity gel consisting of pepstatin-aminobutyl-agarose was prepared for the isolation of these high-molecular-weight renins. The purification of these two high-molecular-weight species were accomplished by the combination of affinity chromatography and three additional steps of conventional chromatography. Electrophoretically homogeneous big renin was found to have a specific activity approximately 1/5 that of the low-molecular-weight (40,000) renin. Two peaks of big-big renin also determined to be electrophoretically homogeneous were found to possess very low specific activities, 3 orders of magnitude lower than the most active renin. In view of the inverse relationship between the molecular weight and the specific activity of big-big, big, and small renin, it has been postulated that a precursor-enzyme relationship represented by the following sequence exists:

big-big renin ⟶ big-renin ⟶ renin

ACKNOWLEDGMENT

We are greatly indebted to Dr. Hamao Umezawa for his kind gift of pepstatin and to Dr. Mizoguchi for inhibitor peptides of renin. This research has been supported by research grants from National Institutes of Health HL-14192 and HL-16114 and grants-in-aid from the American Heart Association 76-124.

REFERENCES

1. Oparil, S., and Haber, E. (1974) N. Eng. J. Med. 291, 389-401
2. Davis, J. O. (1973) Am. J. Med. 55, 339-350
3. Zanchetti, A., Stella, A., Leonetti, G., Morganti, A., and Terzoli, L. (1976) Am. J. Cardiol. 37, 675-691
4. Laragh, J., and Sealey, J. E. (1973) in Handbook of Physiology (Orloff, J., and Berliner, R. W., eds) Section 8 Renal Physiol., pp. 831-908, American Physiological Society, Washington, D.C.
5. Werle, E., Vogel, R., and Goldell, L. F. (1957) Arch. Exp. Pathol. 230, 236-244
6. Koch, C., and Unger, H. J. (1969) Naunyn-Schmiedeberg Arch. Pharmakol. 264, 257-258
7. Cohen, S., Taylor, J. M., Murakami, K., Michelakis, A. M., and Inagami, T. (1972) Biochemistry 11, 4286-4292
8. Haas, E., Lumfrom, H., and Goldblatt, H. (1953) Arch. Biochem. Biophys. 42, 368-386
9. Passanti, G. T. (1959) Biochim. Biophys. Acta 34, 246-248
10. Nairn, R. C., Chadwick, C. S., and Fraser, K. B. (1960) Brit. J. Exp. Path. 41, 214-221
11. Maier, G. D., and Morgan, W. S. (1966) Biochim. Biophys. Acta 128, 193-195

12. Peart, W. S., Lloyd, A. M., Thatcher, G. N., Lever, A. F.,
    Payne, N., and Stone, N.(1966) Biochem. J. 99, 708-716
13. Skeggs, L. T., Lentz, K. E., Kahn, J. R., and Hochstrasser, H.
    (1967) Cir. Res. 21, Suppl. II, 91-107
14. Newsome, H. H. (1969) Biochim. Biophys. Acta 185, 247-250
15. Waldhäusl, W. K., Lucas, C. P., Conn, J. W., Lutz, J. H.,   and
    Cohen, E. L. (1970) Biochim. Biophys. Acta 221, 536-548
16. Lucas, C. P., Fukuchi, S., Conn, J. W., Berlinger, F. G.,
    Waldhäusl, W. K., Cohen, E. L., and Rovner, D. R. (1970)
    J. Lab. Clin. Med. 76, 689-700
17. Poulsen, K., Burton, J., and Haber, E. (1975) Biochim. Biophys.
    Acta 400, 258-262
18. Lauritzen, M., Damsgaard, J. J., Rubin, I., and Lauritzen, E.
    (1976) Biochem. J. 155, 317-323
19. Devaux, C., Menard, J., Sicard, P., and Corvol, P. (1976)
    Eur. J. Biochem. 64, 621-627
20. Majestravich, Jr., J., Ontjes, D. A., and Roberts, J. C. (1974)
    Proc. Soc. Exp. Biol. Med. 146, 674-679
21. Burton, J., Poulsen, K., and Haber, E. (1975) Biochemistry 14,
    3892-3898
22. Umezawa, H., Aoyagi, T., Morishima, H., Matsuzaki, M., Hamada,
    H., and Takeuchi, T. (1970) J. Antibiot. 23, 259-262
23. Aoyagi, T., Morishima, H., Nishizawa, R., Kunimoto, S., Takuchi,
    T., and Umezawa, H. (1972) J. Antibiot. 25, 689-694
24. Gross, F., Lazar, J., and Orth, H. (1972) Science 175, 656
25. Miller, R. P., Pope, C. J., Wilson, C. W., and Devito, E. (1972)
    Biochem. Pharmacol. 21, 2941-2944
26. McKown, M. M., Workman, R. J., and Gregerman, R. I. (1974)
    J. Biol. Chem. 249, 7770-7774
27. Corvol, P., Devaux, C., and Menard, J. (1973) FEBS Lett. 34,
    189-192
28. Murakami, K., Inagami, T., Michelakis, A. M., Cohen, S. (1973)
    Biochem. Biophys. Res. Commun. 54, 482-487
29. Murakami, K., and Inagami, T. (1975) Biochem. Biophys. Res.
    Commun. 62, 757-763
30. Rubin, I. (1972) Scand. J. Clin. Lab. Invest. 29, 51-58
31. Matoba, T., Murakami, K., and Inagami, T. to be published
32. Murakami, K., Inagami, T., and Haas, E., Cir. Res. in press
33. Pickens, P. T., Bumpus, F. M., Lloyd, A. M., Smeby, R. R., and
    Page, I. H. (1965) Cir. Res. 17, 438-448
34. Reniharz, A., Roth, M., Haefeli, L., and Schaechtelin, G. (1971)
    Enzyme 12, 212-218
35. Delpierre, G. R., and Fruton, J. S. (1965) Proc. Natl. Acad.
    Sci. U.S.A. 54, 1161-1167
36. Rajagopalan, T. G., Stein, W. H., and Moore, S. (1966) J. Biol.
    Chem. 241, 4295-4297
37. Tang, J. (1971) J. Biol. Chem. 246, 4510-4517
38. Inagami, T., Misono, K., and Michelakis, A. M.,(1974) Biochem.
    Biophys. Res. Commun. 56, 503-509

39. McKown, M., and Gregerman, R. I. (1975) Lif Sci. 16, 71-79
40. Workman, R. J., and Inagami, T. (1975) Endocrinology 95 Suppl., 138
41. Bayliss, R. S., Knowles, J. R., Wybrandt, G. B. (1969) Biochem. J. 113, 377-386
42. Keilova, H. (1970) FEBS Lett. 6, 312-314
43. Sodek, J., and Hofmann, T. (1970) Can. J. Biochem. 48, 1014-1016
44. Roth, M., and Reinharz, A. (1966) Helv. Chim. Acta 49, 1903-1907
45. Skeggs, L. T., Lentz, K. E., Kahn, J. R., and Schumway, N. P. (1958) J. Exp. Med. 108, 283-297
46. Skeggs, L. T., Lentz, K. E., Kahn, J. R., and Hochstrasser, H. (1968) J. Exp. Med. 128, 13-34
47. Suketa, Y., and Inagami, T. (1975) Biochemistry 14, 3188-3194
48. Morris, B. J., and Lumbers, E. R. (1972) Biochim. Biophys. Acta 289, 385-391
49. Boyd, G. W. (1973) in Hypertension '72 (Genet, J., and Koiw, E., eds) pp. 161-169, Springer, Berlin and New York
50. Leckie, B. (1973) Clin. Sci. 44, 301-304
51. Boyd, G. W. (1974) Cir. Res. 35, 426-438
52. Day, R. P., and Luetscher, J. A. (1975) J. Clin. Endocrinol. Metab. 38, 923-926
53. Day, R. P., and Leutscher, J. A. (1975) J. Clin. Endocrinol. Metab. 40, 1085-1093
54. Skinner, S. L., Cran, E. J., Gibson, R., Taylor, R., Walter, W. A. W., and Catt, K. J. (1975) Am. J. Obstet. Gynecol. 121, 623-630
55. Leckie, B. J., and McConnell, A. (1975) Cir. Res. 36, 513-519
56. Inagami, T., and Murakami, K. (1975) Circulation 52 Suppl. II, 14
57. Levine, M., Lentz, K. E., Kahn, J. R., Dorer, F. E., Skeggs, L. T. (1976) Cir. Res. 38 Suppl. II 90-94
58. Barrett, J. D., Eggena, P., Sambhi, M. P. (1977) Cir. Res. in press
59. Aoi, W., Grim, C. E., and Weinberger, M. H. (1977) Cir. Res. in press
60. Inagami, T., Murakami, K. (1977) Cir. Res. in press
61. Day, R. P., Leutscher, J. A., and Gonzales, C. M. (1975) J. Clin. Endocrinol. Metab. 40, 1078-1084
62. Slater, E. E., and Haber, E. (1976) Circulation 54, Suppl. II 143 Abs
63. Reniharz, A., and Roth, M. (1969) Eur. J. Biochem. 7, 334-339

INACTIVE RENIN - A RENIN PROENZYME?

B. J. Leckie, A. McConnell, and J. Jordan

M. R. C. Blood Pressure Unit
Western Infirmary
Glasgow, G11 6NT

Renin cleaves an α-globulin substrate to produce the decapeptide angiotensin I. This is converted by other enzymes to the vasoconstrictor octapeptide angiotensin II. Renin is produced mainly by the kidney although extra-renal sources of the enzyme have been reported (1,2). Boyd (3) showed that in pig kidneys there is a slow-acting renin with a molecular weight (m.w.) of 60,000 as against 40,000 for active pig renin. After acidification, the slow-acting renin becomes more active and its m.w. is 40,000. In rabbit kidney extracts (4) there is an inactive renin of m.w. around 55,000, which, after exposure to pH 3.0, becomes active and its m.w. falls to 37,000, similar to that of active renin. Morris and Johnston (5) have reported an acid-activatable inactive renin of m.w. 44,000 in rat kidneys. In human kidney extracts an acid-activatable renin has been detected (6) and Day and Leutscher (7) and Day, Leutscher, and Gonzales (8) showed that inactive renin of m.w. 60,000 is present in some renal tumors and extracts of kidneys from diabetic patients. Eggena, Barrett, Silpipat and Sambhi (9) have shown that a high m.w. renin is present in human kidney. Slater and Haber (10) isolated a renin of m.w. 63,000 from normal human kidney. An acid-activatable renin is also present in human amniotic fluid (11) and human plasma (12,13,14).

The inactive forms of renin could be precursors of active renin by analogy with many other proteolytic enzymes which are synthesized as high-molecular-weight inactive proenzymes or zymogens and then undergo a limited proteolysis to produce the active enzyme. Some peptide hormones such as insulin (15) also have large inactive precursors. This paper describes the evidence for the presence of inactive renin in humans and rabbits and discusses the nature of the material.

THE INACTIVE RENIN OF HUMAN PLASMA

Activation by acid

Skinner, Lumbers, and Symonds (16) showed that plasma from
pregnant women dialyzed to pH 3.3 prior to assay at pH 7.5 had a
higher concentration of renin than non-acidified plasma. They
suggested that this was due to the presence of acid-activatable inactive renin in maternal plasma.

To detect the presence of inactive renin in normal plasma, we
dialyzed samples of plasma to various levels of pH at $4^\circ C$, as previously described (13) (Fig.1). The samples were dialyzed back to
pH 7.0 against an assay buffer containing the angiotensinase inhibitors, EDTA and o-phenanthroline (17), and assayed for renin by
measuring the rate of production of angiotensin from ox renin substrate. The substrate was calibrated with standard renin from the
Biological Standards Division of the Medical Research Council. This
renin is fully active (13). Results are expressed as standard international units (18); 190 of our previous units are equivalent to one
standard international unit. Figure 2 shows that plasma exposed to
pH 3.0 had a higher renin concentration than plasma dialyzed at
pH 4 or above. Recovery of the product of the reaction, angiotensin
I, in the incubation mixture was over 90% both for plasma dialyzed
at pH 3.0 and plasma dialyzed at neutral pH, so the difference in
renin concentration was not due to a faster destruction of angiotensin
by nonacidified plasma. The increase in renin concentration after
acidification occurred in all of 11 normal male plasma samples and
8 normal female samples (13). Plasma from pregnant women showed a
proportionally greater increase in renin concentration after acidification, as Skinner, Cran, Gibson, Taylor, Walters, and Catt (19) have
reported. The concentration of inactive renin in plasma can be
calculated by subtracting the renin concentration in plasma dialyzed
at pH 7.4 (active renin) from the renin concentration in plasma
dialyzed at pH 3.0 (total renin):

Total renin = Active renin + Inactive renin
(pH 3.0)   (pH 7.4)

Inactive renin = Total renin - Active renin
(pH 3.0)   (pH 7.4)

% Inactive renin = $\dfrac{\text{Inactive renin}}{\text{Total renin}}$ x 100

Activation by trypsin

Cooper, Osmond, Scaiff, and Ross (20) and Murray and Osmond (21)
have shown that treatment of normal plasma with trypsin results in an

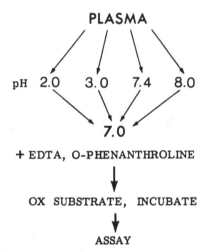

Figure 1. Scheme for the assay of renin in plasma after dialysis to various levels of pH.

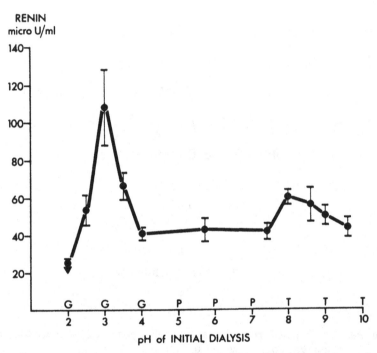

Figure 2. Mean and standard error of renin concentration in
five normal plasma samples after predialysis to pH levels
between 2 and 10.

G = Glycine/HCl buffer   P = Sodium phosphate buffer
T = Tris/HCl buffer

increase in renin activity. Morris and Lumbers (22) showed that pepsin and trypsin activate renin in human amniotic fluid. We tested the effect of trypsin on plasma to see if renin concentration increased.

Samples (1.0 ml each) of a pool of pregnancy plasma were shaken at $37^{\circ}C$ in polythene test-tubes with 0.1 ml of a suspension of Enzite-trypsin, an insoluble trypsin coupled to agarose (Miles Laboratories Ltd.). At various intervals the Enzite-trypsin was removed by centrifugation, 0.5 ml of the supernatant plasma was removed, dialysed against pH 7.0 assay-buffer and assayed renin. Figure 3 shows that pretreatment with trypsin resulted in an increase in renin concentration. If pH 7.4 phosphate buffer was incubated with Enzite-trypsin, the trypsin-free supernate assayed, no renin activity was present. This indicates that the increase in renin concentration after treatment with trypsin was not due to an effect of residual trypsin on the renin assay system. The pH of plasma after 24 hr incubation with trypsin was 7.6. Loss of renin activity occurred in the plasma incubated at $37^{\circ}C$ without trypsin. To see if this loss was due to instability of active renin at $37^{\circ}C$ or to a possible inhibitory effect of plasma on renin (24), partially purified active kidney renin was incubated at $37^{\circ}C$ in the pH 7.4 phosphate-saline buffer (4) that contained neomycin sulphate. After 24 hr, the recovery of renin activity was 48 ± 8%. Renin incubated for the same time at $37^{\circ}C$, pH 7.4 in the presence of normal human plasma gave 62 ± 4% recovery. When acidified pregnancy plasma was incubated at $37^{\circ}C$, the recovery of endogenous renin was 60 ± 5% after 24 hr, as against a recovery of 50 ± 6% for renin in nonacidified plasma. Thus the loss seems due to instability of active renin over 24 hr under the conditions of the incubation at $37^{\circ}C$. Renin is stable at $4^{\circ}C$, pH 7.4 as we have reported (13).

As a further control, the Enzite-trypsin was incubated with soybean trypsin inhibitor (SBTI) prior to incubation with plasma. One mg or 3 mg of SBTI in pH 7.4 phosphate buffer was added to the 0.1 ml of Enzite-trypsin suspension, and the mixture stood for 15 min before 1.0 ml of plasma was added. Figure 4 shows that the presence of SBTI prevented the activation of renin by trypsin. If 4 mg of SBTI was added to standard renin and the mixture then incubated with ox substrate, the SBTI had no effect on the renin-substrate reaction.

Figure 5 shows the renin concentration in plasma incubated for 3 hr with various amounts of Enzite-trypsin. The increase in renin concentration depends on the amount of trypsin present.

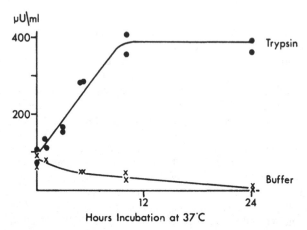

Figure 3. Renin concentration in pregnancy plasma preincubated with Enzite-trypsin for various times at 37°C before removal of the trypsin and assay for renin.

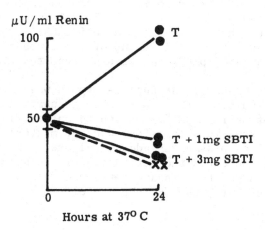

Figure 4. Renin concentration after 24 hr preincubation of plasma with trypsin (T) or trypsin and soybean trypsin inhibitor (SBTI). x---x plasma incubated without trypsin.

Table I

Renin activity in plasma treated with
acid or with trypsin

Pretreatment	Renin Activity	
	Plasma 1	Plasma 2
	μU/ml (mean ± S.E.)	
Dialysis to pH 3.0, 4°C	302 ± 15	121 ± 20
Trypsin 37°C	351 ± 15	113 ± 11
Dialysis to pH 7.4, 4°C	73 ± 14	51 ± 7
No trypsin 37°C	30 ± 4	26 ± 0.2

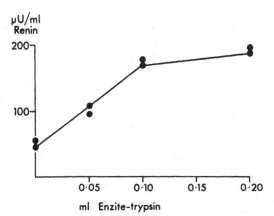

Figure 5. Renin concentration in plasma incubated for 3 hr
with various amounts of Enzite-trypsin.

Comparison of trypsin and acid-activation

Two pools of pregnancy plasma were acidified to pH 3.0 and the renin concentration compared with the results obtained when the plasma was treated with trypsin. Table I shows that there is no difference between trypsin and acid-activation. It is possible that trypsin activates the same inactive renin that acid does (13). To test this, plasma was acidified to pH 3.0 and then dialyzed back to pH 7.4. A sample was assayed for renin and the remainder of the plasma was incubated with trypsin and then assayed again. Table II shows that the renin concentration in the acidified plasma was not significantly different from that in plasma which had been both acidified and treated with trypsin. The renin in nonacidified plasma showed the expected increase in concentration after trypsin treatment.

In another experiment the opposite sequence was tested. Plasma was incubated with trypsin, the trypsin was removed, a sample of the plasma was acidified and assayed for renin. Renin was also assayed in a sample of the trypsin-treated plasma that had been dialyzed to pH 7.4 instead of being acidified. Table III shows that acidification produces no further increase in the renin concentration of trypsin-treated plasma. These results indicate that acid and trypsin act on the same inactive renin.

Sephadex G-100 chromatography of plasma

If the increase in renin concentration that occurs after treatment of plasma by acid or by trypsin is due to activation of a large inactive form of renin, the inactive renin may be separated from the active renin by gel chromatography (4). Twenty-ml of pregnancy plasma was applied to a 2.8 x 90-cm column of Sephadex G-100 and eluted with pH 7.4 saline-phosphate buffer. Four-ml fractions were collected and assayed for renin after being treated as follows: A portion of each fraction was dialyzed at pH 7.4 and assayed. Figure 6a shows that renin was present mainly in fractions 43-47. Another portion of each fraction was acidified to pH 3.0, dialyzed back to pH 7.4 and assayed. Renin concentration increased in fractions 40-44 (Fig.6b) as might be expected if these fractions contained a high-molecular-weight renin that was activated after acidification. A third portion of each fraction was incubated with trypsin. Figure 6c shows that trypsin produced an increase in renin concentration in the same fractions that acidification did. This indicates that these fractions contained an acid-activatable inactive form of renin.

Fractions containing inactive renin were pooled and a portion of the pool was chromatographed again on a 1.5 x 28-cm column of Sephadex G-100 (Fig.7). Inactive renin had a m.w. of around 55,000, with

Table II

Renin activity of plasma after various sequencial
treatment of acid and trypsin

Pretreatment of plasma	Renin activity
	µU/ml (mean ± S.E.)
Dialysis to pH 3.0 then trypsin at 37°C	144 ± 14
Dialysis to pH 3.0	132 ± 13
Dialysis to pH 7.4 then trypsin at 37°C	137 ± 0
Dialysis to pH 7.4	53 ± 4

Table III

The effect of trypsin and dialysis upon the
activation of renin activity in plasma

Pretreatment of plasma	Renin activity
	µU/ml (mean ± S.E.)
Trypsin + dialysis at pH 3.0	326 ± 16
Trypsin + dialysis at pH 7.4	330 ± 10
Dialysis to pH 3.0 alone	302 ± 15
Dialysis to pH 7.4 alone	73 ± 14
Trypsin alone	351 ± 15

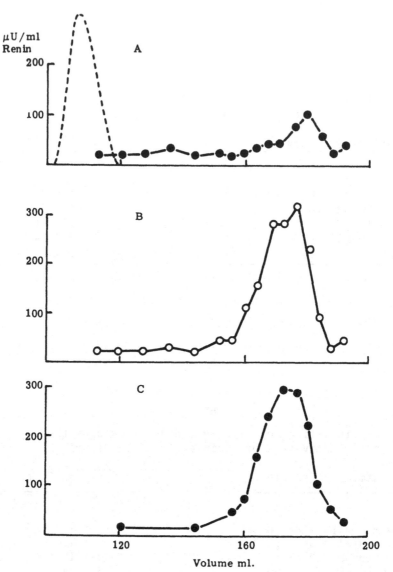

Figure 6. Sephadex G-100 chromatography of pregnancy plasma renin. Fractions were assayed for renin at pH 7.0:
    a) After dialysis at pH 7.4.
    b) After dialysis at pH 3.0.
    c) After incubation with Enzite-trypsin for 10 hr

--- Void volume (Blue Dextran).

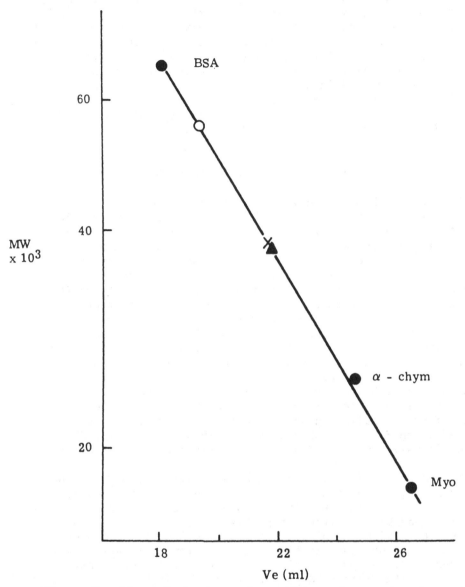

Figure 7. Plot of elution volume (Ve) against molecular
weight for standard proteins bovine serum albumin (BSA)
α-chymotrypsinogen (α-chym) and myoglobin (myo). Fractions
containing inactive renin were pooled from the column shown
in Figure 6 and a portion eluted (0). Inactive renin was
also acidified and then re-run (▲). Standard renin was
also run (X).

reference to a calibration curve made with bovine serum albumin,
α-chymotrypsinogen, myoglobin and Blue Dextran after the method
of Andrews (25). Another portion of inactive renin was acidified
and chromatographed again. Acidified inactive renin had a m.w. of
37,000 to 38,000; therefore the increase in renin concentration that
occurs after acidification is associated with a fall in m.w. Standard
active human kidney renin was eluted in the same fractions as
acidified inactive renin and active human plasma renin. In six
experiments the m.w. of human inactive renin was 54,000 ± 4,000
(mean ± S.E.) and of active renin 40,000 ± 2,000.

THE INACTIVE RENIN OF HUMAN KIDNEY

Extracts of human kidney were assayed to see if they also
contained inactive renin. However, the extracts showed peptidase
activity that was not inhibited by the EDTA and o-phenanthroline
buffer. To measure renin, the kidney extracts were chromatographed
on DEAE-cellulose to remove excess peptidases (angiotensinases),
before being assayed. Kidneys were ground up with sand and phosphate buffer, pH 7.4, in a ratio of 5 ml per gram of cortex (6).
The mixture was centrifuged. A portion of the supernatant was
assayed for protein, and the remainder was equilibrated with
0.005 M sodium phosphate buffer, pH 7.0. Five ml of supernatant
was applied to a 1 x 22-cm column of DEAE-cellulose in the same
buffer. The column was washed with 150 ml of the 0.005 M phosphate buffer, pH 7.0, and renin was eluted with 100 ml of 0.2 M
sodium phosphate buffer, pH 6.0, (4). The effluent was dialyzed
against distilled water, freeze-dried, and the renin taken up in
2.5 ml of pH 7.4 saline phosphate buffer. The solution was
dialyzed to pH 3.0 or to pH 7.4 as described for plasma (13),
diluted in assay buffer and assayed for renin and angiotensinase.

The recovery of angiotensin incubated with a 50-fold dilution
of five kidney preparations before and after extraction on DEAE-
cellulose is shown in Table IV. Before extraction, kidney preparations that had been dialyzed at pH 7.4 destroyed about 50% of the
angiotensin in 24 hr at 37°C. After extraction, good recovery of
angiotensin was obtained in kidney preparations that were dialyzed
at either pH 7.4 or pH 3.0 prior to assay. Renin could therefore
be assayed in the extracted preparations. Table V shows active and
inactive renin per mg of cortical tissue in eight cadaver kidneys
from people who had no record of renal hypertension. There was
considerable variation in the renin content of the kidneys which
was not due to variations in the extraction procedure, as parallel
extractions of the same kidney showed (Table V). Recovery of active
and inactive renin through the extraction was first measured using
amniotic fluid which contained high levels of active and inactive
renin but little angiotensinase (11). The active and inactive renin

Table IV

Effect of predialysis (to pH 3 or pH 7) on the
recovery of angiotensin in kidney extracts

The incubation mixture contained 0.5 ml of extract and
2 ml of 100 ng/ml solution of angiotensin I, as previously
described (13). Removal of angiotensinases was carried out
by chromatography of kidney extracts on DEAE-cellulose.

	pH 3.0	pH 7.4	p
	mean ± S.E.[a]		
Before chromatography	100 ± 5	53 ± 7	<0.01
After chromatography	94 ± 5	85 ± 6	n.s.

[a]Mean and S.E. for five kidney extracts.

Table V

Active and inactive renin in extracts from eight
kidneys after DEAE-cellulose chromatography

Kidney No.	Active renin	Inactive renin	% Inactive renin
	μU renin/mg cortex		
1	41	17	30
2	96	78	45
3	355	166	32
4	30	0	0
5	19	0	0
6	10	7	41
7	12	6	37
8	300	142	32
Repeated extractions (n = 4)			
8 Mean ± S.E.	410 ± 29	142 ± 11	26 ± 1

could therefore be measured before and after extraction on DEAE-
cellulose and the recovery estimated. Recovery of active renin in
three experiments (39 ± 3%) was similar to that of inactive renin
(40 ± 8%). The recovery of active and inactive renin from kidney
extracts that had been chromatographed once and were rechromatograph-
ed was similar: 41% active and 39% inactive.

These results show that there is an inactive form of renin in
human plasma that has a m.w. of around 54,000 as against 37,000 for
active human plasma renin. After acidification or treatment with
trypsin, the inactive renin becomes active and the m.w. falls to
around 37,000. In normal people, the inactive renin comprises 56%
of the total detectable in plasma after acidification to pH 3.0 (13).
An acid-activatable renin was present in six out of eight human
kidney extracts. The percentage varied from 30 to 45% of the total
renin content (mean ± SE = 36 ± 3). This is lower than the percentage of inactive renin in plasma, and together with the fact that
inactive renin is present in anephric patients (13) may indicate
that the material is not derived from the kidney. The percentage of
inactive renin in cadaver kidneys may be slightly different from
that found in viable kidneys. There is circumstantial evidence to
show that the inactive renin of normal humans is derived from the
kidney. Firstly, in the six kidneys which contained inactive renin,
the concentration of both active and inactive renin per mg of extracted protein was high compared with the amount of renin per mg
protein in plasma. For instance, kidney No.7 (Table VI) contained
in a mg protein 213 μU of active and 131 μU of inactive renin. The
concentration of active renin in normal plasma is 0.73 μU/mg protein
and of inactive renin 0.96 μU/mg protein. Secondly, inactive renin
is present in renal vein plasma (12). High levels of active
(563 ± 73 μU/mg) and inactive (929 ± 282 μU/mg: 60% inactive) renin
were present in a kidney with a stenosis of the renal artery from a
patient with raised plasma angiotensin II concentration. Removal of
the kidney lowered the plasma angiotensin II concentration to normal
in this patient.

The inactive renin in humans may be a material similar to the
inactive renin of rabbit kidney extracts.

THE INACTIVE RENIN OF RABBIT KIDNEY

In rabbit kidney extracts there is an inactive renin of m.w.
around 55,000, which after acidification to pH 2.5 becomes more
active and the m.w. falls to 37,000 (4). The inactive material may
also be activated if kidney extracts are chromatographed on DEAE-
cellulose with a shallow gradient from 0.075 M pH 7.0 to 0.3 M pH 6.0
sodium phosphate buffer (4). Active renin is eluted first, then a
material which, when incubated with active renin for 30 min at 37°C,
inactivates it.

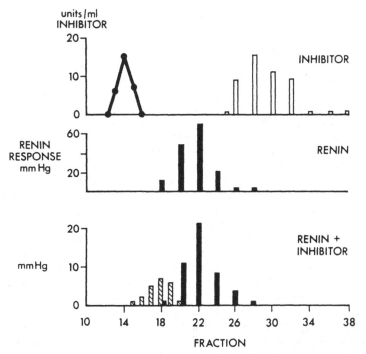

Figure 8. Elution from Sephadex G-100 of rabbit renin inhibitor material ▭ , active rabbit renin ▬ , and a mixture of active rabbit renin and inhibitor ▨ material incubated for 30 min at 37°C.

The activity of the mixture of enzyme and inhibitor can be restored by acidification. Since the inhibitor material is destroyed by acidification to pH 2.5, the inactive renin of rabbit kidney seems to be a complex of active renin and an acid-labile renin inhibitor. Figure 8 shows the elution from Sephadex G-100 of active renin, renin inhibitor, and a mixture of active renin and inhibitor that had been incubated at 37°C for one hr (4). This is similar to the results reported by Boyd for porcine kidney (3). The inactive human renin may similarly be a renin-inhibitor complex.

DISCUSSION

The experiments described above show that a high-molecular-weight inactive form of renin is present in the human and the rabbit. It may be similar to the large forms of renin reported by other workers (3,5,7,8,9,10,12). As yet we have no evidence to prove that the inactive renin is a proenzyme precursor of the active enzyme. If it is a proenzyme analagous to trypsinogen and pepsinogen, the activation step should involve a limited proteolysis, as outlined in Figure 9. In fact the inactive human renin is activated by trypsin, which also activates other prohormones and proenzymes through a limited proteolysis. The inactive renin of the rabbit (4) and the pig (3), however, can be activated by DEAE-cellulose chromatography with removal of an inhibitor (4) or binding protein (3). This suggests that inactive renin is a complex of active renin with an inhibitor that is non-covalently bound to the active enzyme (26,27). The rabbit renin inhibitor is destroyed by acid (4) as is porcine renin binding protein (3). Activation of inactive renin may involve destruction of the bound inhibitor by acid and by trypsin rather than a limited proteolysis. In this case, the inactive renin can hardly be classed as a proenzyme, since the inhibitor may be synthesized entirely independently of renin, and the two may only later become associated, as outlined in Figure 10. A third possibility is that the active renin and the inhibitor found in rabbit kidney are both derived from a very high-molecular-weight precursor. This may undergo a limited proteolysis as the initial activation step, to produce an acid-labile peptide that acts as an inhibitor of active renin. For instance, the activation of pepsinogen results under certain conditions in the formation of pepsin-inhibitory peptides (28). Very high molecular weight precursors of insulin have been isolated (15). This possibility is outlined in Figure 11. In fact, an inactive renin of m.w. 140,000 has been isolated from porcine kidney by Murakami, Matoba and Inagami (29).

Further purification and characterization of the human and rabbit inactive renin is required to determine the biochemical nature of its activation.

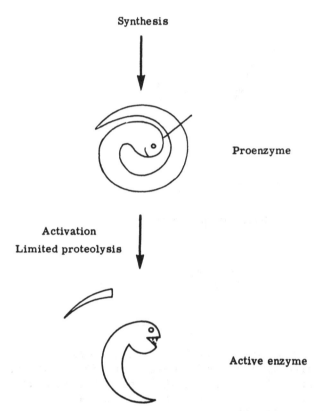

Figure 9. Scheme for activation of a proenzyme by limited proteolysis.

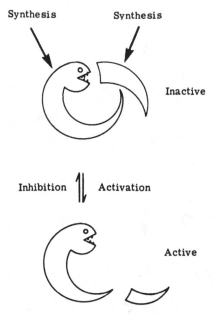

Figure 10. Association and dissociation between an enzyme
and an inhibitor. Inactive rabbit renin can be activated
by removal of an inhibitor on DEAE-cellulose chromatography
or acidification, and active rabbit renin can be inactivated
by incubation with inhibitor.

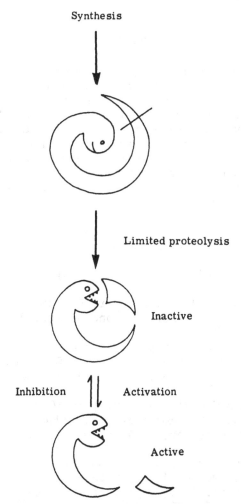

Figure 11. Generation of active enzyme and peptide fragment by limited proteolysis of a proenzyme. The peptide fragment may function as a dissociable inhibitor.

ACKNOWLEDGMENT

Figure 8 is reproduced from an article that appeared in Circulation Research 36, pp. 513: 'A renin inhibitor from Rabbit Kidney', Brenda J. Leckie and Anne McConnell, and is reproduced by permission of Circulation Research.

SUMMARY

Human plasma contains an inactive form of renin with a m.w. of 55,000, as against around 40,000 for active human renin. After acidification to pH 3.0 or incubation with trypsin, the inactive renin becomes more active and the molecular weight falls to that of active renin. In human kidney extracts, an increase in renin concentration also occurred after acidification. The inactive form of renin may be similar to that of rabbit kidneys, which contain an inactive renin that can be activated by removal of an acid-labile inhibitor. The question as to whether the inactive material is a renin proenzyme is discussed.

REFERENCES

1. Carretero, O. A., and Houle, J. A. (1970) Amer. J. Physiol. 218, 689-692
2. Ganten, D., Minnich, J. L., Granger, P., Hayduk, K., Brecht, H. M., Barbeau, A., Boucher, R., and Genest, J. (1971) Science 173, 64-65
3. Boyd, G. W. (1974) Circ. Res. 35, 426-438
4. Leckie, B. J., and McConnell, A. (1975) Circ. Res. 36, 513-519
5. Morris, B. J., and Johnston, C. I. (1976) Endocrinology 98, 1466-1474
6. Leckie, B., and McConnell, A. (1975) in 10th Acta Endocrinologica Congress, pp. 280, Amsterdam, Periodica, Copenhagen
7. Day, R. P., and Luetscher, J. A. (1974) J. Clin. Endocrinol. Metab. 38, 923-926
8. Day, R. P., Luetscher, J. A., and Gonzales, C. M. (1975) J. Clin. Endocrinol. Metab. 40, 1078-1093
9. Eggena, P., Barrett, J. D., Silpipat, C., and Sambhi, M. P. (1976) Fed. Proc. 35, 705
10. Slater, E. E., and Haber, E. (1976) Circulation 53 and 54 (Suppl. II), 143
11. Lumbers, E. R. (1971) Enzymologia 40, 329-336
12. Derkx, F. H. M., Gool, J. M. G., Wenting, G. J., Verhoeven, R. P., Man in't Veld, A. J., Schalekamp, M. A. D. H. (1976) Lancet II, 496-498
13. Leckie, B. J., McConnell, A., Grant, J., Morton, J. J., Tree, M., and Brown, J. J. (1977) Circ. Res. 40 (In Press)

14. Leckie, B., Brown, J. J., Lever, A. F., McConnell, A., Morton, J. J., Robertson, J. I. S., Tree, M. (1976) Lancet II, 748

15. Yalow, R. S. (1974) Recent Progress in Hormone Research 30, 597-633

16. Skinner, S. L., Lumbers, E. R., and Symonds, E. M. (1972) Clin. Sci. 42, 479-488

17. Tree, M. J. (1973) J. Endocrinol. 56, 159A-171

18. Bangham, D. R., Robertson, I., Robertson, J. I. S., Robinson, C. J., and Tree, M. (1975) Clin. Sci. and Mol. Med. 48, 135s-159s

19. Skinner, S. L., Cran, E. J., Gibson, R., Taylor, R., Walters, V. A. W., and Catt, K. J. (1975) Amer. J. Obstet. Gynecol. 121, 626-630

20. Cooper, R. M., Osmond, D. H., Scaiff, K. D., and Ross, L. J. (1974) Fed. Proc. 33, 584

21. Murray, G. E., and Osmond, D. H. (1975) Fed. Proc. 34, 367

22. Morris, B. J., and Lumbers, E. R. (1972) Biochim. Biophys. Acta 289, 385-391

23. Schwert, G. W., and Takenaka, Y. (1953) Biochim. Biophys. Acta 16, 570-575

24. Regoli, D. (1970) Can. J. Physiol. Pharmacol. 48, 350-358

25. Andrews, P. (1964) Biochem. J. 91, 222-233

26. Kunitz, M., and Northrop, J. H. (1936) J. Gen. Physiol. 19, 991-1007

27. Saklatvala, J., Wood, G. C., and White, D. D. (1976) Biochem. J. 157, 339-351

28. Van Vanukis, H., and Herriott, R. M. (1956) Biochim. Biophys. Acta 22, 537-543

29. Murakami, K., Matoba, T., and Inagami, T. (1976) Fed. Proc. 35, 1355

CHARACTERISTICS AND FUNCTIONS OF PROTEINASE A

AND ITS INHIBITORS IN YEAST

Helmut Holzer, Peter Bünning, and Franz Meussdoerffer

Biochemisches Institut der Universität Freiburg im Breisgau
and Institut für Biochemie der Gesellschaft für Strahlen- und
Umweltforschung, Freiburg im Breisgau, Germany

Proteolytic enzymes in yeast have long been known (Willstätter and Grassmann (1). As early as 1917, K. G. Dernby (2) described three types of proteolytic enzymes in extracts and autolysates of yeast. More recently, four different proteinases have been characterized and purified from yeast extracts; some of their properties are summarized in Table I. In this paper, recent work on the "acid" proteinase A and its specific inhibitors from yeast is described as well as possible biological functions of this proteinase A-inhibitor system.

PROTEINASE A

Proteinase A was purified from yeast extracts, activated by incubation at pH 5, by repeated chromatography on DEAE-Sephadex followed by alcohol precipitation, as described by Hata et al. (3). The enzyme could be further purified in our laboratory by affinity chromatography of the partially purified enzyme, using the specific proteinase A-inhibitor I^A, bound to activated CH-Sepharose 4B as the stationary phase. The pattern of elution by 6 M urea from the affinity column is shown in Figure 1; the peaks of hemoglobin-hydrolyzing activity (i.e. proteinase A activity) and protein concentration coincide. Table II summarizes the present purification procedure for proteinase A. Work on further purification and characterization of the enzyme is in progress.

Proteinase A is usually classified as an "acid proteinase". In fact, at pH 3 the enzyme rapidly hydrolyzes acid-denatured hemoglobin, a standard substrate for acid proteinases. At pH values higher than 3, the specific activity of the proteinase decreases (5,6).

Table I

Intracellular proteinases from Saccharomyces cerevisiae

Proteinase	Characteristic	Molecular weight	Specific inhibitors from yeast	Intracellular localization
Proteinase A	Acid endoproteinase (carboxyl proteinase?) glycoprotein	30-45,000	I^A_2 , I^A_3	Vacuoles
Proteinase B	Serine- and cysteine-endoproteinase	30-35,000	I^B_1 , I^B_2	Vacuoles
Carboxypeptidase Y	Serine- and cysteine-exoproteinase, glyco-protein	61,000	I^C	Vacuoles
Aminopeptidase I	Metallo exoproteinase (Zn^{++})	200,000	?	Vacuoles

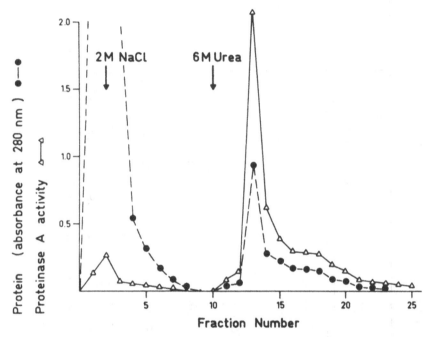

Figure 1. Affinity chromatography of proteinase A. Proteinase A activity was measured with hemoglobin as substrate at pH 3 according to Saheki and Holzer (4).

Table II

Purification of proteinase A

Purification procedure	Total activity units[a]	Specific activity units . mg^{-1}	Yield %	Purification folds
Crude extract	< 2,000	< 0.3	-	-
Activation at pH 5 by proteolysis of I^A	440,000	7.6	100	1
Ammonium sulfate precipitation	335,000	43.5	76	1.5
DEAE-Sephadex A-50 chromatography	52,000	130	12	25
Affinity chromatography [b]	34,500	2000	9	260

[a]Units of proteinase A activity were defined as described by Saheki and Holzer (4).

[b]Affinity chromatography runs were carried out with 7,500 units of enzyme from the DEAE-Sephadex A-50 chromatography step.

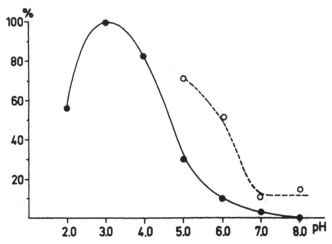

Figure 2. Proteolysis of acid-denatured hemoglobin and inactivation of apotryptophan synthase by proteinase A as a
function of pH.
(●) Hydrolysis of hemoglobin measured according to Saheki and
Holzer (4) in % of the value at pH 3,
(o) % inactivation of apotryptophan synthase. Assay as described by Katsunuma et al. (8).

Physiological substrates, however, such as apotryptophan synthase,
cytoplasmic malate dehydrogenase or NAD-dependent glutamate dehydrogenase from yeast, are attacked by proteinase A at neutral pH (4,7).
Results from experiments on the sensitivity of hemoglobin and yeast-
apotryptophan synthase, respectively, to yeast proteinase A are
shown in Figure 2.

PROTEINASE A-INHIBITORS

In the last two years inhibitors for the proteinases A and B,
and carboxypeptidase Y have been isolated from yeast (6,9-14); some
of their properties are summarized in Table III. Two subgroups of
inhibitors for the proteinases A and B, respectively, have been
separated according to differences in electrophoretic mobility, i.e.
in isoelectric point (see Table III). In early experiments (6),
evidence for four different species of proteinase A-inhibitors had
been obtained, but only the species I^A2 and I^A3 could be further
purified and characterized (6,11). It has been shown recently
(Bünning and Holzer: in press) that in contrast to Baker's yeast,
which contains the two inhibitors for proteinase B (10), in several
strains of Saccharomyces cerevisiae only I^B2 could be demonstrated
and in Saccharomyces carlsbergensis, only I^B1. Single cell cultures
from Baker's yeast contain I^B1 : I^B2 in the same concentration ratio
of 1:3 as observed in preparations from commercial Baker's yeast.
The mixture of inhibitors is therefore not the result of a mixture
of yeast strains in the commercial yeast. Furthermore, because the
restriction of I^B1 to Saccharomyces carlsbergensis and of I^B2 to
Saccharomyces cerevisiae has been found at all growth conditions
studied, it may be concluded that they are coded by two different
genes, only one of which is contained in Saccharomyces cerevisiae
and Saccharomyces carlsbergensis, respectively; both genes are
present in Baker's yeast. These two different inhibitors, I^B1 and
I^B2, therefore may justifiably be called "isoinhibitors" in analogy
to "isoenzymes". We assume that the "isoinhibitors" is also applica-
ble to I^A2 and I^A3. Investigation of this situation is in progress.

The amino acid compositions of the heat-stable inhibitors I^B2
and I^A3 are shown in Table IV. Neither inhibitor contains arginine,
tryptophan and cysteine. I^B2 also lacks methionine, I^A3 proline.
Both inhibitors obviously contain a relatively high percentage of
polar amino acids (lysine, glutamic acid + glutamine, aspartic acid +
asparagine). From the amino acid composition the minimal molecular
weight has been calculated to be 8,500 for I^B2 and 7,700 for I^A3.
Under all conditions studied only a monomeric form of I^B2 has been
found. In contrast, Sephadex G-75 gel filtration of I^A3 indicated
an equilibrium of a monomer with an oligomer, molecular weight of
23,000. Recent experiments with I^A3 in sedimentation and equilibrium
runs in the analytical ultracentrifuge, however, gave no evidence for
an aggregation of the monomer. Typical results are shown in Figure 3.

Table III

Proteinase-inhibiting polypeptides from yeast

Inhibitor	Isoelectric point	Specific inhibition of	Hydrolyzed by	Molecular weight	Stability	Intracellular localization
I^A_2	5.7	Proteinase A	Proteinase B	7,700	Heat- and acid resis- tant	Cytosol
I^A_3	6.3					
I^B_1	8.0	Proteinase B	Proteinase A	8,500	Heat- and acid-resis- tant	Cytosol
I^B_2	7.0					
I^C	6.6	Carboxypeptidase Y	Proteinases A and B	23,800	Heat- and acid-labile	Cytosol

Table IV

Amino acid composition of $I^B 2$ and $I^A 3$

| Amino acid | Nearest integer | |
	$I^B 2$ [a]	$I^A 3$ [b]
	residues/molecule	
Asx	12	10
Thr	5	3
Ser	2	6
Glx	7	13
Pro	2	0
Gly	3	5
Ala	2	5
Val	8	2
Ile	5	1
Leu	6	2
Tyr	1	2
Phe	3	2
Lys	11	13
His	6	2
Arg	0	0
Met	0	2
Cys	0	0
Trp	0	0
Sum:	73	68

[a] The results of $I^B 2$ are part of the Ph.D. thesis of K. Maier, University of Freiburg.

[b] The result of $I^A 3$ are taken from Núñez de Castro and Holzer (11).

The linear relationship of the logarithm of optical density to the square of the radius indicates the presence of only one type of particle. Using a measured partial specific volume of v_2^* = 0.72 cm^3/g, a molecular weight of 6,500 was calculated from the slope of the curve shown in Figure 3. Taking into account the considerable inaccuracy involved when determining molecular weights below 10,000 using ultracentrifuge methods, this is reasonably close to the molecular weight of 7,700 calculated from the amino acid composition, the latter being probably the true molecular weight. We now assume that during filtration on Sephadex G-75, the appearance of protein peaks corresponding to a molecular weight of about 23,000 results from a deviation of the molecules from globular form.

Figure 4 describes data from Saheki et al. (6) concerning the pH and concentration-dependence of proteinase A inhibition by I^A3. At high concentration ratios of inhibitor to proteinase A, inhibition does not exceed 60% at pH 3, whereas at pH 7 almost complete inhibition is attained. Assuming reversibility of the pH-dependent changes in inhibition, this means that at high ratios of inhibitor to proteinase A, a change in pH from 3 to 7 causes an increase in inhibition, whereas acidification from pH 7 to 3 causes deinhibition, i.e. activation or proteinase A in a previously inhibited incubation mixture. A possible biological role of this phenomenon in the control of proteinase A activity is discussed in the following section.

POSSIBLE BIOLOGICAL FUNCTIONS OF PROTEINASE A AND PROTEINASE A-INHIBITORS

Table V summarizes three types of proteolytic processes which may be of biological significance in yeast cells. It has been found that proteinases A and B, and carboxypeptidase Y as well as aminopeptidase I (15) are localized in the vacuoles of yeast cells; this supports the hypothesis that selected enzymes are degraded by uptake into the vacuoles followed by hydrolysis by cooperative action of the four proteinases. In the hypothetical scheme shown in Figure 5 (16) it is assumed that the selection for degradation is mediated by signals reflecting changes in the physiological conditions which make the enzymes or other proteins vulnerable for uptake by the vacuoles. The susceptibility for this uptake might be caused by an effector-stimulated change in conformation or by an enzyme-catalyzed covalent interconversion. Catabolite inactivation in yeast, such as the glucose-induced rapid disappearance of certain enzyme or permease activities (for a review see Holzer {16}), is probably an example of such a selective proteolytic degradation of proteins dependent on a change in the metabolic conditions. Results from Neeff and Mecke (personal communication) demonstrated very recently a coincidence of the disappearance of cytoplasmic malate dehydrogenase activity and of the serological reactivity of this enzyme with a specific antiserum after the addition of glucose to acetate-grown yeast cells.

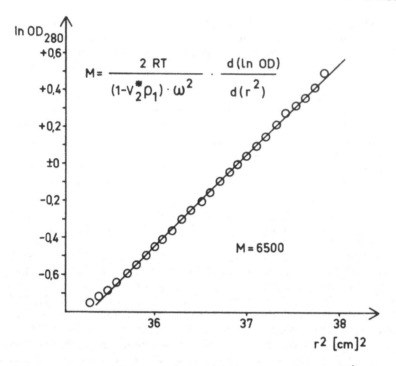

Figure 3. Sedimentation diffusion equilbrium of I^A3 as measured in the analytical ultracentrifuge. A molecular weight of 6,500 was calculated using the equation shown in the figure.

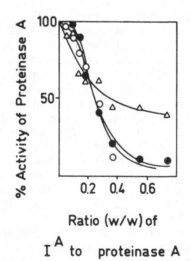

Figure 4. Inhibition of proteinase A by I^A3 at pH 3.0 ($\triangle$), 5.0 ($\bullet$) and 7.0 (o).

Table V

Proteolytic processes in yeast

Process	Mechanism	Site of proteolysis	Possible example	Role of specific intracellular inhibitors
Degradation of enzymes	Enzyme to be degraded enters vacuole	Inside vacuole	Catabolite inactivation; starvation; transition to spore formation	Protection in extravacuolar space against unwanted lysis of vacuoles
Activation of inactive proenzymes by limited proteolysis	Small vacuoles release proteinases	Outside vacuole (cell, membrane, etc.)	Activation of chitin synthase	Restriction of proteolysis to site of chitin synthase activation
Formation of subunits by limited proteolysis of precursor proteins	Proteinase cuts precursor protein	Mitochondrion	Assembly of cytochrome oxidase from cytosolic precursor protein and mitochondrial subunits	Restriction of proteolysis to site of cytochrome oxidase assembly

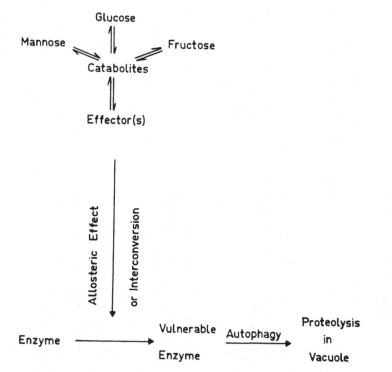

Figure 5.   Hypothetical scheme for catabolite inactivation
in yeast.

These findings support the idea of a proteolytic mechanism for catabolite inactivation. The extremely high sensitivity of cytosolic
malate dehydrogenase, in contrast to its mitochondrial isoenzyme,
against proteolysis by the yeast proteinases A and B (7) also
supports this idea.

The data in Table VI demonstrate that the proteolytic activities
in yeast at the stationary phase are about ten times higher than in
the earlier log phase of growth. A similar relationship is found
for the inhibitory activities against proteinases A and B. Probably
the biologic function of increased proteolytic activity at starvation is to degrade enzymes which are no longer necessary after
growth stops. Free amino acids are thus freed for other purposes,
for instance biosynthesis of enzymes necessary for uptake and metabolism of substrates in a new growth cycle. The increase of cytosolic proteinase inhibitors parallel to proteinase activity (see
Table VI) might point to the inhibitors as "safety guards" against
unwanted proteolysis in the extravacuolar space: the more proteinases
are accumulated in the vacuoles, the more protection in their surroundings may be necessary.

Another situation where existing enzymes become unnecessary and
new enzymes must be synthesized occurs at the transition from vegetative growth to spore formation. The experiments by Betz and
Weiser (18) depicted in Figure 6 demonstrate that in a yeast suspension kept in a sporulation medium, proteinase A and proteinase
A-inhibitors increase considerably hours before appearance of the
mass of spores. These changes and the parallel increase in proteinase B observed by these investigators are evidence in support of
the postulated function of proteinases in degradation of superfluous
enzymes and of inhibitors as "safety guards" against high concentrations of proteinases. Good evidence for participation of
proteinase A in the transition from vegetative growth to spore
formation is provided also by the finding (19) that a yeast mutant
low in proteinase A activity does not sporulate.

In Table VII it is shown that the activities of proteinase A
and its inhibitors are low in yeast growing on glucose as the carbon
source as compared to yeast growing on acetate. From these data and
from other experiments (20), we assume that the mechanism causing
regulatory increase of the proteinase A and I^A- activities at
starvation and sporulation is a release from catabolite repression
by glucose.

An example of an extravacuolar proteolytic process precisely
controlled in space and time is the activation of chitin synthase
by limited proteolysis (21,22,14). Chitin synthase catalyzes
synthesis of the chitin, which is part of the septum between mother
and daughter cell in budding yeast, and of the budscar remaining
after separation of the daughter cell.

Table VI

Proteinase activities and inhibitory activities against proteinase A and B in crude extracts from yeast grown in batch culture (for details see {17})

| Culture time | Cell density | Proteinase A | | Inhibitory activity against proteinase A[b] | Proteinase B after activation[c] | Inhibitory activity against proteinase B[e] | Proteinase C after activation[d] |
| | | Fresh crude extract | After activation[a] | | | | |
	A 550 nm	units/mg	units/mg	units/mg	A 520 nm/min/mg	A 520 nm/min/mg	μmol/min/mg
9	0.70	0.51	0.96 (42)	0.45	0.0055	0.018	0.031
16	4.3	1.2	2.6 (52)	1.4	0.019	0.065	0.055
24	5.3	1.3	3.3 (23)	2.0	0.039	0.12	0.095
48	7.3	3.1	7.9 (40)	4.8	0.061	0.15	0.14

[a] Hours of activation in brackets.

[b] Calculated as difference "after activation" minus "fresh crude extract" with the assumption that activation is a measure of the inhibitor hydrolyzed at incubation.

[c] Activity in fresh crude extract was less than 2% of the activity measured after activation.

[d] The increase of proteinase C activity at activation was variable and ranged between 100 and 500% of the initial activity.

[e] Assayed in boiled extracts.

Table VII

Proteinase A and proteinase A-inhibitor (I^A) in yeast cells
growing with and without glucose and harvested at early log phase
(for details see {20})

	Specific activity (Units/mg supernatant protein)	
	Glucose-medium	Acetate-medium
Proteinase A	0.87	2.43
I^A	0.023	0.11

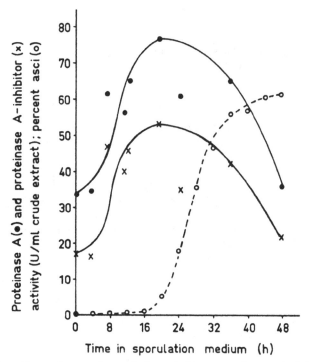

Figure 6. Proteinase A activity (•), proteinase A-inhibitor
level (x) and percentage asci (o) during sporulation.

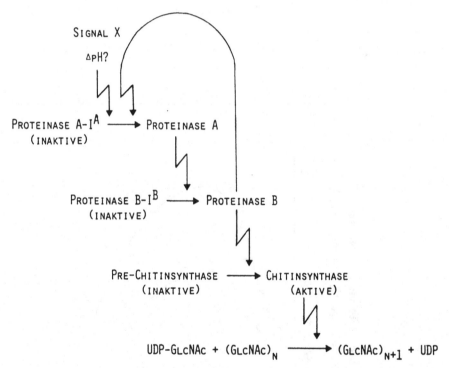

Figure 7. Cascade mechanism of activation of chitin synthesis in yeast.

The activating proteinase, and the corresponding specific heat-stable inhibitor, which can prevent activation, discovered by Cabib et al. (14), were shown to be identical with the previously described proteinase B and the proteinase B-inhibitor, I^B, respectively (23). The finding (10,17) that proteinase A hydrolyzes and thereby inactivates the proteinase B-inhibitor, even when the inhibitor is bound to proteinase B, suggests a cascade mechanism of activation of chitin synthase as depicted in Figure 7 (24). The signal initiating the cascade might be a local decrease in pH; this could activate inhibited proteinase A as discussed above in connection with the experiments shown in Figure 4. Proteinase A activates inhibited proteinase B, and this active proteinase B can then activate chitin synthase. The activation of inhibited proteinase A by proteinase B, which hydrolyzes I^A (6,17), suggests a feedback loop in this cascade which would amplify the proteolytic activation of proteinase B and thereby speed up the whole cascade. Each of the single steps used for construction of the cascade shown in Figure 7 have been shown to function in vitro, but the functioning of the complete system in vivo has not been demonstrated; it may serve as a working hypothesis for further studies.

Yeast cytochrome oxidase is built up from seven different subunits, of which four are synthesized in the cytosol, and three in the mitochondria (25). R. O. Poyton (26) recently reported evidence suggesting that the four cytosolic subunits are synthesized in one peptide chain. This precursor protein might be split by limited proteolysis in connection with or after transport from the cytosol to the inner mitochondrial membrane. Here seems to be an example of a proteolytic process in yeast participating in subcellular translocation of subunits and assembly of an enzyme (see Table V). Which proteinase is involved in this process, and whether specific inhibitors also play a part, is as present unknown.

ACKNOWLEDGMENTS

The authors are indebted to Dr. E.-G. Afting and Mr. H.-P. Henninger for the ultracentrifugation studies and to Dr. J. S. Elce for discussion and a critical reading of the manuscript. The assistance of Miss D. Montfort is gratefully acknowledged.

SUMMARY

A purification and some properties of proteinase A from yeast are described. A specific macromolecular inhibitor of proteinase A from yeast cytosol has been isolated and shown to be a protein (molecular weight 7,700) consisting of a majority of polar amino acids. Proline, arginine, cysteine and tryptophan were not detected in the inhibitor. Possible biological functions of proteinase A and the proteinase A-inhibitor (and of other yeast proteinases and their inhibitors) in the following processes are discussed: general protein turnover, catabolite inactivation of enzymes, enzyme degradation at starvation and at transition to spore formation, and activation of pre-enzymes and precursor proteins by limited proteolysis.

REFERENCES

1. Willstätter, R., and Grassmann, W. (1926) Hoppe-Seyler's Z. Physiol. Chem. 153, 250-282
2. Dernby, K. G. (1917) Biochem. Z. 81, 107-208
3. Hata, T., Hayashi, R., and Doi, E. (1967) Agric. Biol. Chem. 31, 357-367
4. Saheki, T., and Holzer, H. (1974) Eur. J. Biochem. 42, 621-626
5. Hata, T., Hayashi, R., and Doi, E. (1967) Agric. Biol. Chem. 31, 150-159
6. Saheki, T., Matsuda, Y., and Holzer, H. (1974) Eur. J. Biochem. 47, 325-332
7. Jusiĉ, M., Hinze, H., and Holzer, H. (1976) Hoppe-Seyler's Z. Physiol. Chem. 357, 735-740
8. Katsunuma, T., Schött, E., Elasässer, S., and Holzer, H. (1972) Eur. J. Biochem. 27, 520-526

9. Ferguson, A. R., Katsunuma, T., Betz, H., and Holzer, H.
    (1973) Eur. J. Biochem. 32, 444-450
10. Betz, H., Hinze, H., and Holzer, H. (1974) J. Biol. Chem. 249,
    4515-4521
11. Núñez de Castro, I., and Holzer, H. (1976) Hoppe-Seyler's Z.
    Physiol. Chem. 357, 727-734
12. Matern, H., Hoffmann, M., and Holzer, H. (1974) Proc. Natl.
    Acad. Sci. U.S.A. 71, 4874-4878
13. Lenney, J. F. (1975) J. Bacteriol. 122, 1265-1273
14. Ulane, R., and Cabib, E. (1974) J. Biol. Chem. 249, 3418-3422
15. Matile, P., and Wiemken, A. (1967) Arch. Microbiol. 56, 148-155
16. Holzer, H. (1976) Trends in Biochem. Sci. 1, 178-181
17. Saheki, T., and Holzer, H. (1975) Biochim. Biophys. Acta 384,
    203-214
18. Betz, H., and Weiser, U. (1976) Eur. J. Biochem. 62, 65-76
19. Betz, H. (1976) Proc. of the Symposium on Cell Differentiation
    in Microorganisms, Plants and Animals, Reinhardsbrunn (April
    (1976) Deutsche Akademie der Naturforscher, Leopoldina (in press)
20. Hansen, R. J., Switzer, R. L., Hinze, H., and Holzer, H. (1976)
    Biochim. Biophys. Acta (in press)
21. Cabib, E., and Ulane, R. (1973) Biochem. Biophys. Res. Commun.
    50, 186-191
22. Cabib, E., Ulane, R., and Bowers, B. (1973) J. Biol. Chem. 248,
    1451-1458
23. Hasilik, A., and Holzer, H. (1973) Biochem. Biophys. Res.
    Commun. 53, 552-559
24. Holzer, H., and Saheki, T. (1976) Tokai J. Exp. Clin. Med. 1,
    115-125
25. Schatz, G., and Mason, T. L. (1974) Annu. Rev. Biochem. 43, 51-87
26. Poyton, R. O. (1977) in Genetics and Biogenesis of Mitochondria
    and Chloroplasts (Bücher, T., Werner, S., and Neupert, W., eds)
    North Holland Elsevier (in press)

HUMAN CATHEPSIN D

Alan J. Barrett

Strangeways Research Laboratory
Cambridge, England

Our interest in cathepsin D arose from the observation that
agents such as vitamin A alcohol and complement-sufficient antiserum,
which caused a dramatic catabolism of cartilage matrix in a culture
system, simultaneously caused a sharp increase in the synthesis and
secretion of cathepsin D by cells in the tissue. Clearly this was
consistent with the possibility that the acid proteinase was responsible for the degradation of the cartilage matrix, and if this
were true for the organ culture system, it might apply to cartilage
degradation in arthritis, too.

PURIFICATION

To find out more about the possible role of cathepsin D in the
pathological tissue damage of rheumatoid arthritis, we want to work
with the chick embryo organ culture system, with an experimental
arthritis model in rabbits, and with human tissue samples; so we
purified and studied the enzyme from each of the three species (1-3).

The purification of cathepsin D has been reviewed in detail
(3,4). In our own work the starting material was liver that had been
stored frozen. The tissue was thoroughly homogenized, and the soluble
extract incubated overnight at acid pH. During this autolysis step,
many contaminating proteins were denatured and/or digested by
cathepsin D, so that subsequent fractionation with cold acetone gave
an excellent purification factor. Ion-exchange chromatography
brought the preparation to approximately 80% purity. The final step
of purification was preparative isoelectric focusing, which resolved
three major forms of human and chicken cathepsin D, but many more for
rabbit. The results of analytical isoelectric focusing of samples

from the purification of the human enzyme are shown in Figure 1.
The three major forms of the enzyme were designated α, β and γ;
their approximate isoelectric points were 5.7, 6.0 and 6.5.

The β and γ forms of human cathepsin D showed a consistent amino
acid composition in several preparations, showed only the characteristic banding patterns in isoelectric focusing or SDS-gel electrophoresis (see below), and gave single precipitin lines in immunodiffusion
against polyvalent antisera raised against partially purified preparations. Moreover, the molarity of active sites titratable with
pepstatin agreed with that predicted from the protein concentration
of the solutions, and the molecular weight determined by gel chromatography (see below). These preparations were therefore judged to
be pure.

STRUCTURE AND COMPOSITION

The enzyme (γ-form) was hydrolyzed for amino acid analysis by
standard methods, with or without performic acid oxidation. The
results of the analyses (Table I) were corrected for losses of the
more labile residues. Probably each molecule contains four disulphide bonds since thiol groups have not been detected. One or more
of the disulphide bonds seems to be important in stabilizing the
active conformation of cathepsin D. At pH 8 this bond may be
strained, as it is reducible by 10^{-3}M dithiothreitol, and not re-
formed following exposure to oxygen (4). The presence of glucosamine shows that cathepsin D is a glycoprotein, but nothing more is
known about the carbohydrate component.

It is now clear that the molecular weight of cathepsin D from
a variety of sources is close to 42,000 (3-7). The elution volume
in gel chromatography is typically identical with that of hen egg
albumin, and it is unclear why higher values (50,000 - 60,000) were
obtained in some of the earlier work.

In SDS-gel electrophoresis, cathepsin D may run as a single
polypeptide chain, or as fragments representing about one-third and
two-thirds of the molecule, respectively. Sapolsky and Woessner (8)
found that the majority of forms of bovine cathepsin D give the two
components, although a few do not. The major forms of human cathepsin D give only the split products of molecular weight about 27,000
and 14,000, but each component appears as a double band (Fig.2).
Our tentative conclusion is that the protein is synthesized as a
single chain which may well fold as a large and a smaller globular
subunit, the two being linked by a segment rather susceptible to
proteolysis, and which is cleaved in one of two points at some stage
before we get it on to the SDS-gel electrophoresis. In the native
molecule, even after cleavage, the two parts of the molecule are

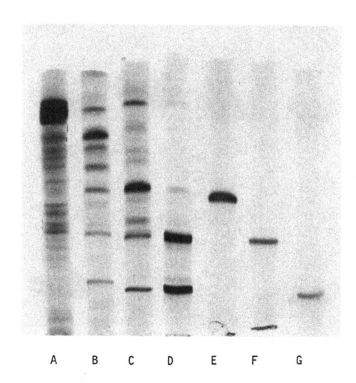

Figure 1. Analytical Isoelectric Focusing at Various Stages
in the Purification of Human Cathepsin D. (a) Homogenate,
(b) after autolysis and acetone fractionation, (c) after DEAE-
cellulose chromatography, (d) after CM-cellulose chromatography,
(e) α-form, (f) β-form and (g) γ-form, after preparative
isoelectric focusing (from Ref. 2).

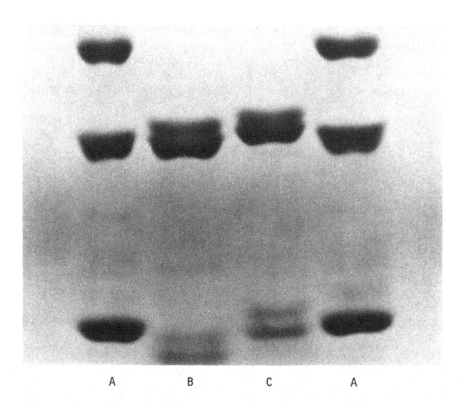

Figure 2. The polypeptide chains of human cathepsin D. The
samples run in electrophoresis in the presence of sodium
dodecyl sulphate were (a) molecular weight markers, ovalbumin
(43,000), carbonic anhydrase (29,000) and chicken lysozyme
(14,300); (b) β-form of human cathepsin D treated with mercaptoethanol; (c) β-form of human cathepsin D treated with
iodoacetate. The direction of migration was downward, in an
electrophoretic system similar to that of Alvares and
Seikevitz (24). Doublets at about 30,000 and 14,000 molecular
weight indicate that there are two bonds close together in the
molecule susceptible to cleavage. No uncleaved molecules
were detected in the region of 43,000 molecular weight.

Table I

Amino acid composition of human cathepsin D (γ-form)

The composition (residues/mole) is calculated on the basis of a molecular weight of 42,000. The figure for tryptophan is derived from a direct spectrophotometric analysis made by Dr. C. G. Knight.

Amino acid	Residues per mole	Amino acid	Residues per mole
Lysine	25	Glycine	40
Histidine	5	Alanine	20
Arginine	10	Valine	31
Half-cystine	8	Isoleucine	22
Aspartic acid	32	Leucine	35
Methionine	14	Tyrosine	21
Threonine	20	Phenylalanine	15
Serine	26	Tryptophan	4
Glutamic acid	36	Glucosamine	7
Proline	23		

held together by noncovalent forces; they separate in the presence of chaotropic agents without cleavage of disulfide bonds. The separate components give distinct precipitin lines in gel double immunodiffusion against antiserum to cathepsin D (4).

ACTIVITY

Like cathepsin D of other species, the human enzyme is very active in the digestion of hemoglobin, its usual test substrate. Many other proteins do act as substrates, but usually with a much slower formation of products soluble in dilute trichloroacetic acid.

The pH optima of cathepsin D from various sources, acting on hemoglobin, are affected by the stability of the enzyme to acid, since the limit of activity on the acid side is usually attributable to irreversible denaturation. In assays with ten minutes incubation at 45°C, human cathepsin D shows maximal activity at pH 3.5 (3).

Because of our interest in the possible role of cathepsin D in the degradation of cartilage, we have been particularly concerned to learn about its action on the two major components of this tissue, proteoglycan and collagen. The pH optimum for degradation of proteoglycan by cathepsin D is 4.5 - 5.0 (9). Since proteolytic degradation of proteoglycan can be sensitively detected by viscometry, this substrate is a convenient as well as a relevant one with which to test the upper limits of pH for the activity of cathepsin D. In agreement with the report of Woessner (10) for the bovine enzyme, we

found that even with 70μg of isoelectrically focused cathepsin D
per ml of proteoglycan solution, and overnight incubation at 37°C,
no activity was detectable (11). This represents a remarkably
complete loss of activity over two pH units and is consistent with
the idea that a conformational change abolishes the active site of
cathepsin D in the range pH 5 - 7 (see below).

Native collagen appears to be entirely resistant to the action
of human cathepsin D (12).

No work has been done on the action of human cathepsin D on low
molecular weight substrates, but work with the enzyme from other
species strongly suggests that the peptide bonds most susceptible to
hydrolysis will be those involving aromatic amino acid residues. In
the oxidized B chain of insulin, the bonds most susceptible to cathepsin D are Leu-15—Tyr-16, Tyr-16—Leu-17, Phe-24—Phe-25, and Phe-25
—Tyr-26. Little activity is to be expected with peptides of less
than five amino acids.

INHIBITORS

Inhibitors represent one of the most powerful tools with which
to investigate the physiological functions of a proteinase, but the
carboxyl proteinases seemed to have no satisfactory low molecular
weight inhibitors, until pepstatin was discovered. Like the other
carboxyl proteinases, cathepsin D is inactivated by diazoacetyl
amino acid esters in the presence of Cu^{2+}, as well as related
carboxyl-blocking reagents (see reference 4 for review).

In the search for a specific inhibitor suitable for physiological
experiments, Dingle et al. (9) explored the immuno-inhibition of
cathepsin D. Antisera of high titre, containing antibodies to at
least six separate determinants on the enzyme molecule, were capable
of completely inhibiting the action of the enzyme on hemoglobin or
proteoglycan at pH 5 or above. At lower pH values dissociation of
the enzyme-antibody complexes decreased the efficiency of inhibition.

Pepstatin appears to be a group-specific inhibitor of the carboxyl proteinases (13,14). Barrett and Dingle (13) showed that
pepstatin is a tightly-binding inhibitor of human cathepsin D that
reacts with the enzyme in equimolar ratio.

Knight and Barrett (7) have confirmed that pepstatin is a good
titrant for the determination of the molarity of solutions of
cathepsin D and have measured the apparent dissociation constants
(K_D) for the binding of pepstatin, pepstatin methyl ester and pepstatinyl [^{3}H] glycine to cathepsin D. At pH 5 and below, K_D was
5×10^{-10}M, but binding was strongly pH-dependent. Qualitatively,

the looser binding at higher pH values was shown by the ease of separation of each of the inhibitors in gel chromatography, with recovery of active enzyme. The effect was quantitated by use of the pepstatinyl [^{3}H] glycine in equilibrium dialysis. As the pH was raised from 5.0 to 6.4, K_D rose from 5 x 10^{-10}M to 2 x 10^{-6}M. Since the catalytic activity of cathepsin D declines essentially to zero as the pH is raised from 5 to 7, it was suggested that the binding site for substrate and pepstatin is abolished by a conformational change in the enzyme molecule. The data indicated that in biological experiments near neutral pH, larger molar excesses of pepstatin over cathepsin D would be required for efficient inhibition.

PHYSIOLOGICAL ROLE

Much remains to be learned about the physiologic functions of cathepsin D. In living cells the enzyme is normally confined in the lysosomes: membrane-limited organelles that contain a range of about 50 acid hydrolases which are responsible for the bulk of catabolism of cellular components. This lysosomal localization has been demonstrated by biochemical means (15), and by immunofluorescence microscopy (16). Dingle et al. (17) used specific antiserum to cathepsin D to show that the enzyme contributes to the digestion of proteins including hemoglobin, immunoglobin G, and cartilage proteoglycan in the lysosomes of living rabbit macrophages. Dean (18) introduced pepstatin into the cells of perfused rat liver in liposomes, and showed that the degradation of cellular proteins was markedly inhibited. In the lysosomal system, cathepsin D presumably acts synergistically with cathepsin B (which is probably still more important in the degradation of many proteins) and the exopeptidases (19). Dean and Barrett (20) point out that both cathepsin B and D are likely to be much more active in the lysosomal system than they appear in test-tube experiments: it can be calculated that both are present in the lysosomes at approximately 10^{-3}M concentration, equivalent to 45 mg of cathepsin D/ml, whereas 1µg/ml shows a very high activity in a standard assay.

The question of an extracellular role for cathepsin D is still an open one. The association of secretion of lysosomal enzymes, including cathepsin D, from cells, with proteolytic degradation of the extracellular cartilage matrix was first detected biochemically, but was confirmed by immunofluorescence microscopy with culture systems and tissue ex vivo (21). Incubation of living tissue in medium containing antibodies resulted in the precipitation of enzyme-antibody complexes around cells that were engaged in secretion of the enzyme. The complexes were then located by use of a fluorescent second antibody. The demonstration that secretion of cathepsin D into cartilage matrix is associated with degradation of the matrix was obviously consistent with a role for cathepsin D in the degradative process.

Further evidence came from the inhibition of the autolytic degradation of killed cartilage at pH 5-6 by both specific antisera to
cathepsin D (22,9) and pepstatin (17). Knight and Barrett (7)
showed that the characteristics of enzyme inhibition by pepstatin
should allow one to inhibit any extracellular enzyme action in
organ culture systems. To date, it has not been possible to obtain
consistent inhibition of the degradation of the matrix of living
cartilage either by the antisera or pepstatin, however. Further
work is being done in our laboratory with new culture systems, and
with derivatives of pepstatin, in an attempt to obtain a truly
clear-cut answer to the question of whether, despite its requirement
for a slightly acidic medium, cathepsin D does participate in extracellular proteolysis.

In the body of high mammals, extracellular cathepsin D probably
is bound by α_2-macroglobulin, once it reaches the circulation, and
thus inactivated and eliminated. Barrett and Starkey (23) showed
that labeled cathepsin D is bound by α_2-macroglobulin at pH 6, and
that the interaction depends upon the activity of the enzyme, since
it was partially blocked by pepstatin (a weak inhibitor at this pH
value). Complexes of proteinases with α_2-macroglobulin are inactive
against protein substrates, and are rapidly cleared from the circulation (23).

SUMMARY

Cathepsin D was purified from human liver by a procedure involving autolysis, acetone fractionation, and chromatography on ion-
exchange media and organomercurial-sepharose. Multiple forms of the
enzyme were then separated by preparative isoelectric focusing.

The molecular weight of the protein was found to be 43,000.
Its amino acid composition was determined and it was shown to be a
glycoprotein. When treated with sodium dodecyl sulphate or
chaotropic agents (without reduction) all forms of the enzyme tested
gave components of about 28,000 and 14,000 molecular weight.

Specific antisera were raised against the enzyme, and the
characteristics of immunoinhibition were investigated. Immuno-
inhibition of rabbit cathepsin D within living macrophages was
shown to interfere with degradation of some proteins endocytosed by
the cells. The antisera against human and rabbit cathepsin D were
used in immunofluorescent localization of the enzyme in sites of
tissue damage in which cathepsin D might be implicated.

The characteristics of inhibition of human cathepsin D by pepstatin were established. At pH values below 5, K_D values of 5 x
10^{-10}M were determined and pepstatin was shown to be an excellent
titrant for cathepsin D. In the range pH 5-6.4 K_D increased steeply

and it was concluded that the binding site for substrate and inhibitor was abolished by a conformational change in the enzyme molecule in which three protons are lost.

REFERENCES

1. Barrett, A. J. (1967) Biochem. J. 104, 601-608
2. Barrett, A. J. (1970) Biochem. J. 117, 601-607
3. Barrett, A. J. (1971) in Tissue Proteinases (Barrett, A. J. ed) pp. 109-133, North-Holland, Amsterdam
4. Barrett, A. J. (1977) in Proteinases in Mammalian Cells and Tissues (Barrett, A. J. ed)
5. Woessner, J. F., Jr. (1971) in Tissue Proteinases (Barrett, A. J. and Dingle, J. T. eds) pp. 291-311, North-Holland Publishing Company
6. Ferguson, J. B., Andrews, J. R., Voynick, I. M., and Fruton, J. S. (1973) J. Biol. Chem. 248, 6701-6708
7. Knight, C. G., and Barrett, A. J. (1976) Biochem. J. 155, 117-125
8. Sapolsky, A. I., and Woessner, J. F., Jr. (1972) J. Biol. Chem. 247, 2069-2076
9. Dingle, J. T., Barrett, A. J., and Weston, P. D. (1971) Biochem. J. 123, 1-13
10. Woessner, J. F., Jr. (1973) Fed. Proc. Fed. Am. Soc. Ex. Biol. 32, 1485-1488
11. Barrett, A. J. (1975) in Proteases and Biological Control (Reich, E., Rifkin, D. B., and Shaw, E. eds) pp. 467-482, Cold Spring Harbor Laboratory, New York
12. Burleigh, M. C., Barrett, A. J., and Lazarus, G. S. (1974) Biochem. J. 137, 387-398
13. Barrett, A. J., and Dingle, J. T. (1972) Biochem. 127, 439-441
14. Takahashi, K., Chang, W. J., and Ko, J. S. (1974) J. Biochem. (Tokyo) 76, 897-899
15. De Duve, C., Wattiaux, R., and Baundhuin, P. (1962) Adv. Enzymol. 24, 291-358
16. Poole, A. R., Dingle, J. T., and Barrett, A. J. (1972) J. Histochem. Cytochem. 20, 261-265
17. Dingle, J. T., Poole, A. R., Lazarus, G. S., and Barrett, A. J. (1973) J. Exp. Med. 137, 1124-1141
18. Dean, R. T. (1975) Nature, London 257, 414-416
19. Barrett, A. J., and Heath, M. F. (1977) in Lysosomes: A Laboratory Handbook (Dingle, J. T. ed) 2nd Edn. North-Holland, Amsterdam (in press)
20. Dean, R. T., and Barrett, A. J. (1976) in Essays in Biochemistry (Campbell, P. N., and Aldridge, W. N. eds) Vol. 12, pp. 1-40, Academic Press, London
21. Poole, A. R., Hembry, R. M., and Dingle, J. T. (1974) J. Cell Sci. 14, 139-161

22. Weston, P. D., Barrett, A. J., and Dingle, J. T. (1969) Nature London, 222, 285-286

23. Barrett, A. J., and Starkey, P. M. (1973) Biochem. J. 133, 709-724

24. Alvares, A. P., and Siekevitz, P. (1973) Biochem. Biophys. Res. Commun. 54, 923-929

UNIQUE BIOCHEMICAL AND BIOLOGICAL FEATURES OF CATHEPSIN D IN RODENT LYMPHOID TISSUES

William E. Bowers, John Panagides, and Nagasumi Yago

The Rockefeller University, New York, New York 10021

During the course of studies on lysosomes in rat thoracic duct lymphocytes (TDL), it was found that cathepsin D, a typical lysosomal enzyme in most cells and tissues, did not show the same distribution as the other lysosomal acid hydrolases after fractionation. In this paper, we report some of our findings concerning the unique properties of this rat TDL enzyme.

METHODS

The method of obtaining and handling rat thoracic duct lymphocytes (TDL) has been detailed previously (1). The preparation of homogenates of rat TDL, fractionation by differential and isopycnic centrifugation, and calculations were carried out according to procedures described by Bowers (1), Beaufay et al. (2), and Leighton et al. (3). Zonal centrifugation of rat TDL was performed according to the method described previously (4). Enzymes were assayed according to Bowers et al. (5), Yago and Bowers (6), and Tulkens et al. (7). Chromatography on Sephadex G-100 and G-200, partial purification of cathepsin D, inhibition by pepstatin and by antisera have been described previously (6).

RESULTS AND DISCUSSION

Differential centrifugation. Rat TDL are difficult cells to rupture under isotonic conditions, but it can be achieved if they are subjected to hypotonic shock (1) or treated with rabbit anti-rat lymphocyte antiserum and complement. Homogenates of rat TDL prepared by these two methods and fractionated by differential centrifugation

301

yield similar distributions for all lysosomal enzymes except cathepsin D. As seen in Table I, two typical lysosomal enzymes N-acetyl-β-glucosaminidase and β-glucuronidase, contribute 65% of their total activity to a high-speed pellet, whereas cathepsin D, which is measured on denatured bovine hemoglobin at an optimal pH of 3.5 and is an extremely active enzyme in lymphocytes, has a completely different distribution. For lymphocytes broken open under hypotonic conditions, most cathepsin D activity is recovered in the high-speed supernatant; for lymphocytes lysed with antibody and complement, a large part of the total cathepsin D activity associates with the nuclear fraction, even though pyrophosphatase, an enzyme located in the cytosol, is recovered mainly in the high-speed supernatant. These results suggest that cathepsin D is located inside particles, especially since we have been able to rule out the possibility that soluble cathepsin D adsorbs onto sedimentable material recovered in the nuclear fraction. It is clear that the major portion of cathepsin D resides in particles that differ from lysosomes.

Isopycnic centrifugation. If a post-nuclear extract of hypotonically shocked rat TDL is fractionated by means of isopycnic centrifugation in an aqueous sucrose density gradient, all of the sedimentable acid hydrolases, including a small part of cathepsin D, band around a modal density of 1.18 (Fig.1). In agreement with the results presented in Table I, most of the cathepsin D activity is recovered in a soluble, unsedimentable form. The other lysosomal acid hydrolases contribute much less unsedimentable activity.

Isopycnic centrifugation of whole homogenates of rat TDL lysed by antibody and complement lead to distributions for the acid hydrolases similar to those seen in Figure 1, except for cathepsin D. The major portion of this enzyme activity sediments through the gradient along with the nuclei, a finding which also agrees with the distribution found after differential centrifugation (Table I). Thus far we have been unable to dissociate the cathepsin D activity from the nuclei.

Figure 1. Distribution of acid hydrolases and protein after isopycnic centrifugation in sucrose gradients of post-nuclear extracts of rat thoracic duct lymphocytes. The shaded block with a density below 1.10 has an arbitrary density interval of 0.1 and represents the position of the sample layer. The dotted line indicates the histogram obtained if enzyme activity or protein were uniformly distributed. Each histogram shows the average of results with standard deviation, and the number of experiments is given between parentheses.

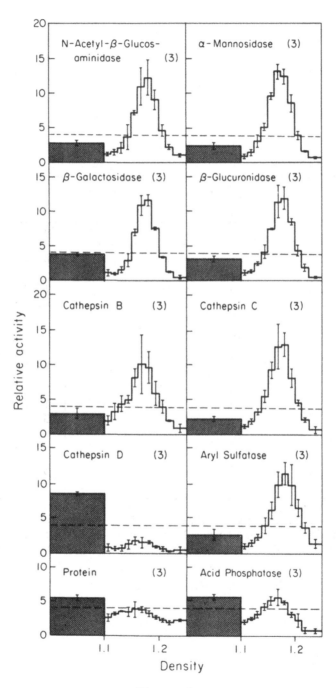

Figure 1

Table I

Fractionation by differential centrifugation of a homogenate
of rat thoracic duct lymphocytes lysed by hypotonic shock
or by antibody and complement

Enzyme	Fraction	% of total activity	
		Hypotonic shock	Antibody-complement
N-Acetyl-β-glucosaminidase	N	14.2 ± 6.6	17.1 ± 4.8
	P	64.6 ± 7.6	63.6 ± 4.8
	S	21.2 ± 8.4	19.3 ± 3.8
β-Glucuronidase	N	13.5 ± 4.3	17.6 ± 4.6
	P	64.5 ± 6.8	64.2 ± 0.4
	S	22.0 ± 5.5	18.2 ± 4.2
Cathepsin D	N	14.6 ± 7.0	52.1 ± 5.6
	P	20.0 ± 7.1	15.2 ± 4.0
	S	65.4 ± 8.0	32.7 ± 4.0
Pyrophosphatase	N	-	7.4 ± 3.8
	P	-	3.1 ± 1.4
	S	-	89.5 ± 4.0

Number of fractionations performed: 8 for hypotonic shock; 4
for antibody-complement. N, nuclear fraction; P, high-speed
pellet fraction; S, high-speed supernatant experiment.

Zonal centrifugation. The existence of cathepsin D in an intracellular location different from that of the lysosomal enzymes suggested that they might reside in two different types of cells. TDL are comprised almost entirely of a homogeneous population of small lymphocytes and a minor population of large lymphocytes which amount to some 1-5% of the total cell preparation. A method of zonal centrifugation was therefore developed to separate TDL mainly according to size. The validity of the method was demonstrated by correlating the size of the cells in each fraction with the distribution of ^{3}H-thymidine, which is incorporated under appropriate conditions almost entirely by large lymphocytes. These results indicated that the large cells associated with the fractions most distal from the axis of rotation. As seen in Figure 2, relatively little of the total enzyme activity for all four acid hydrolases is found in the region occupied by the large cells, and more importantly, the distribution of cathepsin D is quite similar to the lysosomal enzymes. It thus appears that both cathepsin D and the lysosomal enzymes belong mostly to small lymphocytes. Further studies indicated that macrophages do not account for the cathepsin D activity and that both thymus-derived and bone marrow-derived lymphocytes are equally rich in cathepsin D and lysosomal enzymes.

The conclusion that cathepsin D and the other lysosomal enzymes reside together in the same cell is supported by recent findings with mouse leukemia L1210 cells (8). These cells, which presumably represent a homogeneous, clonally-derived population of lymphocytes, also possess normal lysosomes, as well as cathepsin D which occurs mostly in an unsedimentable form after fractionation.

Inhibition by antiserum. An intracellular localization for cathepsin D different from that of the other lysosomal acid hydrolases in rat TDL led us to explore some of the biochemical properties of this enzyme. As illustrated in Figure 3, an antiserum prepared in rabbits against rat liver soluble lysosomal enzymes effectively inhibited rat liver cathepsin D, although it did not inhibit the cathepsin D of rat TDL. In this case the incubations were carried out at pH 5 instead of pH 3.6 to avoid dissociation of the antigen-antibody complex. Both rat liver and rat TDL cathepsin D, however, have identical pH activity curves toward denatured bovine hemoglobin as substrate.

Chromatography on Sephadex G-100. Chromatography of extracts of rat liver and of rat TDL also reveal differences between the enzymes (Fig.4). Two molecular weight forms for rat TDL cathepsin D were found: one, with an apparent molecular weight of 45,000 corresponding to the single form obtained from rat liver, and another having a higher apparent molecular weight of 95,000.

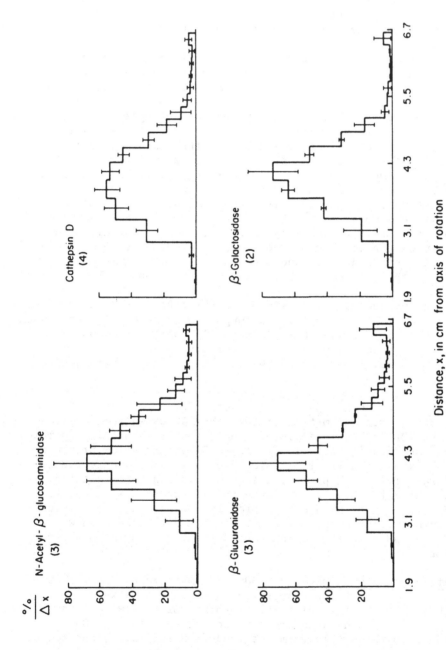

Figure 2. Distribution of acid hydrolases after fractionation of rat thoracic duct lympho-cytes by zonal centrifugation. Histograms were plotted according to Bowers (4). Reproduced with permission from the Journal of Cell Biology.

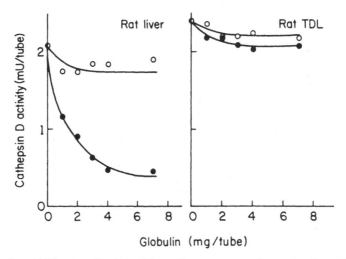

Figure 3. Effect of globulins from sera of control and of
immunized rabbits on cathepsin D activity in high-speed
extracts of rat liver and of rat TDL. High-speed extracts
of rat liver and of rat TDL were preincubated with control
globulins from unimmunized rabbits (o) or globulins from
antisera of rabbits immunized against soluble rat liver lysosomal enzymes (●) for 1 hr at 37°. Enzyme activity was
measured at pH 5.0 according to the description of Yago and
Bowers (5). Reproduced with permission from the Journal of
Biological Chemistry.

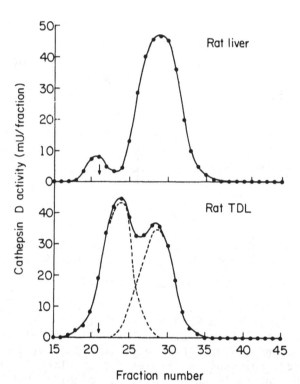

Figure 4. Comparison of the chromatographic behavior of
cathepsin D from rat liver and rat TDL. High-speed extracts
of rat liver and of rat TDL were subjected to chromatography
on Sephadex G-100. The void volume is indicated by the arrow.
The two curves depicted by the dashed lines are those resolved
by a Dupont 310 curve resolver. Reproduced with permission
from the Journal of Biological Chemistry.

Inhibition by pepstatin. The sensitivity to inhibition by pepstatin of both low and high molecular weight rat TDL enzymes differs markedly from that of rat liver cathepsin D, as seen in Figure 5. The lymphocyte enzymes were inhibited to the extent of 5 mU per nanogram of pepstatin, whereas only 1 mU of rat liver cathepsin D was inhibited by one nanogram of pepstatin. Owing to the fact that pepstatin binds in an equimolar ratio to cathepsin D (9), the steeper slope of inhibition found for the two rat TDL cathepsin D in comparison to rat liver cathepsin D indicates that the lymphocyte enzymes have a higher specific activity. Because the low molecular weight form of rat lymphocyte cathepsin D differs in its sensitivity to inhibition by pepstatin, we will refer to it as L-enzyme and to the high molecular weight form as H-enzyme.

No change in the sensitivity of rat liver cathepsin D was noted during the course of a 200-fold purification, and thus it is unlikely that pepstatin-binding contaminants are present in the preparation (6).

Conversion of H-enzyme to L-enzyme. After chromatography on Sephadex G-100, as well as on Sephadex G-200, H-enzyme reproducibly eluted at a position corresponding to an apparent molecular weight of 95,000, slightly more than twice that of the L-enzyme. When a fraction of the pooled preparation of H-enzyme was treated with β-mercaptoethanol, it converted to the low molecular weight form. There was no loss in enzyme activity accompanying this conversion, and recovery from the column was nearly total. Due to its characteristic sensitivity to pepstatin, the low molecular weight enzyme formed after β-mercaptoethanol treatment was found to be L-enzyme (6).

Tissue and species distribution. The existence of unique forms of cathepsin D in rat lymphoid tissues led us to examine other tissues and species for these enzymes. Rat liver cathepsin D is inhibited to exactly the same degree as that obtained from rat kidney, rat adrenal, rat fibroblast, human tonsils, bovine liver, bovine spleen, calf thymus, rabbit liver, and rabbit spleen (6, Fig.5). Thus far, the highly pepstatin-sensitive forms of cathepsin D have been found only in the lymphoid tissues of rodents: rat spleen, thymus, lymph node cells, and lymphocytes, mouse spleen and thymus, a mouse-derived leukemia L1210 line, and hamster spleen.

Classification of H- and L-enzyme as types of cathepsin D. The reasons for classifying H- and L-enzymes as types of cathepsin D are the following:
    1. The pH curves on denatured bovine hemoglobin as substrate are identical to that found for rat liver cathepsin D.
    2. Relative to other denatured protein substrates tested, they are most active on denatured bovine hemoglobin.

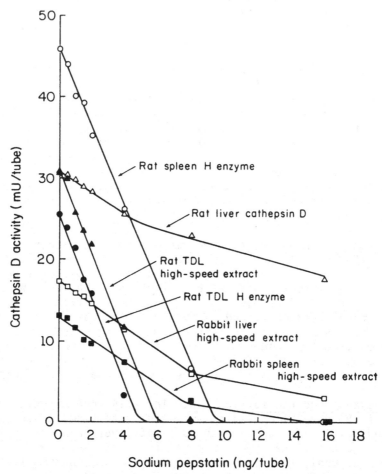

Figure 5. Effect of pepstatin on cathepsin D activities of various tissues in rat and rabbit. Enzyme preparations were pre-incubated with varying amounts of sodium pepstatin. Symbols in the figure are: o, rat spleen H-enzyme (762 mU/mg protein); Δ, rat liver cathepsin D (43,000 mU/mg protein); ▲, rat TDL high-speed extract (103 mU/mg protein); ●, rat TDL H-enzyme (927 mU/mg protein); □ , rabbit liver high-speed extract (18.4 mU/mg protein); and ■ , rabbit spleen high-speed extract (11.7 mU/mg protein). Reproduced with permission from the Journal of Biological Chemistry.

3. Pepstatin, a specific inhibitor of cathepsins D and E, inhibits both H- and L-enzyme, even though their enzyme activity on denatured bovine hemoglobin is greater than that of rat liver cathepsin D.

4. They are not affected by thiol reactive inhibitors, such as TPCK or TLCK, or by leupeptin, an inhibitor of cathepsin B.

5. H-enzyme is not identical to cathepsin E because it hydrolyzes human serum albumin much less effectively than it does bovine hemoglobin; it also converts to L-enzyme after treatment with β-mercaptoethanol.

6. Some of the other known cathepsins, such as A, B_1, and C, are located intracellularly in lysosomes in rat TDL in contrast to cathepsin D which has an uncertain localization in these cells, but which shows only partial, if any, association with lysosomes.

ACKNOWLEDGMENT

It is a pleasure to acknowledge the excellent technical assistance of Ms. Elizabeth Dolci and Ms. Kathy Kohn and the superb typing of Mrs. Anna Polowetzky. This work was supported by Grant IM-67 from the American Cancer Society, Grants HD-05065 and CA-16875 from the U.S. Public Health Services, and Grant GB-35258X from the National Science Foundation.

SUMMARY

Cathepsin D, an enzyme consistently found to be lysosomal in many cells, has an unusual localization in rat thoracic duct lymphocytes (TDL). After fractionation of homogenates of rat TDL, most of the enzyme activity, as measured at pH 3.6 on denatured bovine hemoglobin, is distributed differently from the other lysosomal enzymes. The enzyme also has some unique properties: it is not inhibited by an antiserum inhibitory for rat liver cathepsin D; it exists in two molecular weight forms ($\sim$ 45,000 and $\sim$ 95,000) both of which have a higher specific activity than rat liver cathepsin D, as determined by studies using the irreversible inhibitor, sodium pepstatin; the high molecular weight form converts to the low molecular weight form after treatment with β-mercaptoethanol without any loss in activity. These enzymes appear to be restricted to rodent lymphoid tissues. Reasons for considering them to be a type of cathepsin D are given in the text.

REFERENCES

1. Bowers, W. E. (1972) J. Exp. Med. 136, 1394-1403
2. Beaufay, H., Bendall, D. S., Baudhuin, P., Wattiaux, R., and de Duve, C. (1959) Biochem. J. 73, 628-637

3. Leighton, F., Poole, B., Beaufay, H., Baudhuin, P., Coffey, J. W., Fowler, S., and de Duve, C. (1968) J. Cell Biol. 37, 482-513

4. Bowers, W. E. (1973) J. Cell Biol. 59, 177-184

5. Bowers, W. E., Finkenstaedt, J. T., and de Duve, C. (1967) J. Cell Biol. 32, 325-337

6. Yago, N., and Bowers, W. E. (1975) J. Biol. Chem. 250, 4749-4754

7. Tulkens, P., Beaufay, H., and Trouet, A. (1974) J. Cell Biol. 63, 383-401

8. Bowers, W. E., Beyer, C. F., and Yago, N. (1977) Biochim. Biophys. Acta 497, 272-279

9. Barrett, A. J., and Dingle, J. T. (1972) Biochem. J. 127, 439-441

SPECIFICITY AND BIOLOGICAL ROLE OF CATHEPSIN D

J. Frederick Woessner, Jr.

Departments of Biochemistry and Medicine
University of Miami School of Medicine
Miami, Florida 33152

Cathepsin D was originally isolated from bovine spleen by Press, Porter and Cebra in 1960 (1). My interest in this enzyme was aroused by the observation of its high levels of activity in the involuting post-partum uterus (2); in fact, the uterus contains high levels of this enzyme even in the normal nongravid condition. This led to the purification of cathepsin D from the bovine uterus (3). Uterine cathepsin D was insensitive to a large number of inhibitors specific for thiol, serine, and metallo-proteases, and incapable of cleaving a large group of synthetic peptide substrates typically used for the known proteases (4). Strong inhibition could be obtained only by the use of heavy metals such as Pb^{2+} and Hg^{2+}. The specificity was determined by the use of the S-sulfo B chain of insulin. Most rapid splitting was at the bonds Leu_{15}-Tyr_{16}, Phe_{24}-Phe_{25} and Phe_{25}-Tyr_{26}; secondary splitting was found at Glu_{13}-Ala_{14}, and Tyr_{16}-Leu_{17}. These points of cleavage are the same as reported by Press et al. (1). Minor splits were also observed at Phe_1-Val_2 and Ala_{14}-$\overline{Leu_{15}}$ (4).

In a further study of cathepsin D from bovine uterus, it was shown that as many as 12 distinct isozyme forms could be resolved (5). Six major forms account for 94% of the total activity; these forms were proven to have the same specificity of action on the B chain of insulin. The forms differ in charge and have isoelectric points ranging from pH 5.8 to 7.2. Disc electrophoresis in the presence of sodium dodecyl sulfate and mercaptoethanol demonstrated that the separated forms numbered 2, 3, and 4 gave single bands with mobilities corresponding to molecular weights of 40,000 to 42,000. Forms 5 through 9 dissociated to give two bands each with weights of 13,000 to 14,000, and 25,000 to 28,000. Ten multiple forms of spleen cathepsin D were reported by Press et al. (1).

Barrett (6) noted multiple forms in several species of liver.
Ferguson et al. (7) observed the dissociation from 42,000 to a
28,000 dalton fragment in sodium dodecyl sulfate; but they did not
observe the smaller fragment.

The major single-chain form, form 4, accounts for about 35% of
the total tissue content of cathepsin D. It has a molecular weight
of 40,000 by equilibrium sedimentation. Forms 5 and 6 are dissociated into two chains by sodium dodecyl sulfate; they are also less
stable than form 4 in acid or urea (5). These considerations lead
to the diagram presented in Figure 1. It is suggested that the
original form of the enzyme is the intact chain of molecular weight
40,000 (form 4). This may undergo nicking by some other proteolytic
enzyme to produce form 5. This form has two chains of unequal
length, but their total weight approximates 40,000. Further cleavage
may produce the slightly smaller form 6 by reducing the length of
the larger chain. This does not account for all 12 forms, but
forms 2 and 3 may also represent intact chains that differ slightly
in charge from form 4 and these may also undergo nicking processes.
Interconversions of these forms in vitro have not been possible.
The nicking is likely to occur in the tissue prior to the preparative
procedures.

FURTHER STUDIES OF ENZYME SPECIFICITY

The best preparations of uterine cathepsin D reported to date
had a specific activity of 150 units/mg protein (5). Such preparations had a V_{max} for hemoglobin of 23.5 μequivalents tyrosine re-
leased/min/mg protein. More recently it has been possible to
increase this activity to 200 units/mg by a second chromatography
on DEAE-Sephadex A-25. The enzyme was applied in 5 mM sodium
phosphate buffer, pH 8.4, and washed with 1 l of the same buffer.
The column was then washed with 3 l of 10 mM buffer. This gave a
well-resolved peak of form 4 and a second peak with forms 5 and 12
partially separated. Increasing the buffer strength to 20 mM brought
out a third peak of forms 6 and 7 combined. Form 4 has the highest
specific activity and has been used for specificity studies.

Since cathepsin D does not hydrolyze typical substrates used
for pepsin such as Z-Glu-Tyr or Z-Gly-Phe, it was necessary to design
new peptides that might serve as substrates. Keilova and Keil (8)
synthesized a heptapeptide corresponding to positions 23-29 of the
insulin B chain: Gly-Phe-Phe-Tyr-Thr-Pro-Lys. This was split on
either side of the second Phe residue by cathepsin D from bovine
spleen. In 1971, I reported splitting of the slightly shorter
peptide Glu-Ala-Leu-Tyr-Leu-Val by uterine cathepsin D. This
sequence corresponds to positions 13-18 of the B chain, and the main
point of cleavage is at Leu-Tyr (9).

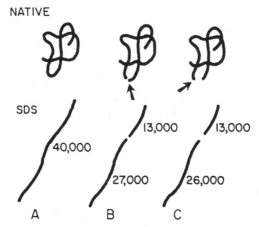

Figure 1. Scheme illustrating the postulated nicking of the single primary chain of cathepsin D. The upper row represents the folded native protein, the lower row represents protein denatured and extended by treatment with sodium dodecyl sulfate and mercaptoethanol. The arrows indicate points of proteolytic cleavage.

Recently two pentapeptides have been prepared with susceptible Phe-Phe sequences and with blocked ends (Table I). With purified cathepsin D, the K_m for the peptide Bz-Arg-Gly-Phe-Phe-Leu-4-methoxy-β-napthylamide is about 3×10^{-4} M, V_{max} is about 450 nmole/min/mg protein and K_{cat} is 0.3 sec^{-1}. Corresponding parameters for the action of pepsin on this peptide are similar, emphasizing the close relationship between these two enzymes. This peptide has a further advantage in that it can be used to localize cathepsin D in histological sections of tissue. The methodology has been described in detail by Smith (10). After enzymatic cleavage of the peptide, aminopeptidase M is added to cleave Phe-Leu from the 4-MBNA moiety, which in turn is coupled with hexazotized pararosanilin to give an insoluble precipitate.

Kinetics for the cleavage of the third peptide (Table I) were similar to those for the one preceding. Cleavage of the remaining peptides was slow, ranging from 5-50% after 18 hours of incubation (15-150 nmole/min/mg enzyme). Phenyl sulfite (also termed diphenyl sulfite) was cleaved by pepsin and cathepsin D at closely comparable rates: about 12 nmole/min/mg enzyme at a substrate concentration of 50 μM. Phenyl sulfite is hydrolyzed by a number of acid proteases (11), but its cleavage by cathepsin D has not been previously reported.

It is interesting to note that four of the peptides in Table I (the 2 MBNA peptides, phenyl sulfite, and Ac-Phe-TyrI$_2$) are split at quite similar rates by cathepsin D and pepsin. This is in striking contrast to a series of seven peptides reported by Sampath-Kumar and Fruton (12) in which pepsin acted 100 to 1,000 times faster than cathepsin D on each peptide.

The cleavage pattern noted for the various peptides in Table I is consistent in requiring strongly hydrophobic residues on either side of the bond to be cleaved. Cathepsin D is able to act as an aminopeptidase on several small Leu- and Phe-peptides. This agrees with the known slow release of Phe from the NH$_2$-terminus of the B chain of insulin and from position 25 in the peptide Phe$_{25}$-Ala$_{30}$. Some Phe was also released from the peptide Phe-Phe-Tyr-Thr (Table I). Leu and Phe were not released from dipeptides which lacked a blocking group on the carboxyl end.

A series of Tyr-containing peptides that were not split by cathepsin D are listed in Table II. Typical pepsin substrates such as Z-Glu-Tyr and Z-His-Tyr-OMe were not cleaved by cathepsin D. It was surprising that Tyr-Leu-NH$_2$ and Tyr-Phe-NH2 were not cleaved in view of the cleavage of the related dipeptides noted in Table I, and the known cleavage of NH$_2$-terminal Tyr from the B chain peptide Tyr$_{16}$-Phe$_{24}$. It has not yet been possible to find a synthetic Tyr-peptide that is cleaved by cathepsin D on the carboxyl side of Tyr.

Table I

Peptides Split By Uterine Cathepsin D

Peptide		Rate
Glu - Ala - Leu -	-Tyr - Leu - Val	++++
Bz - Arg - Gly - Phe -	-Phe - Leu - 4MBNA[a]	+++
Bz - Arg - Pro - Phe -	-Phe - Leu - 4MBNA	+++
Phe - Phe -	-Tyr - Thr	++
Ac - Phe -	-Tyrl_2	++
Leu -	-Tyr - NH_2	++
Leu -	-Phe - NH_2	++
Phe -	-Tyr - NH_2	+
Ala - Leu -	-Tyr - Leu - BNA[a]	+
Ala - Leu -	-Tyr - Leu	+
Phenyl Sulfite		++

[a]-4MBNA is the abbreviation of -4-methoxy-β-naphthylamide; and -BNA is the abbreviation of -β-naphthylamide.

Table II

Tyrosine Peptides <u>Not</u> Split By Cathepsin D

Z - Glu - Tyr	Ac - Gly - Phe - Phe - Tyr - OMe
Gly - Tyr - NH_2	Z - Phe - Phe - Tyr - OMe
Gly - Tyr	Z - Phe - Phe - Tyr
Z - His - Tyr - OMe	Gly - Leu - Tyr
His - Tyr - NH_2	Z - Phe - Tyr - Thr-D-Pro
Z - Leu - Tyr - NH_2	Phe - Tyr - Thr
Leu - Tyr	Z - Tyr - Ala
Ac- Phe - Tyr - NH_2	Z - Tyr - Gly - NH_2
Z - Phe - Tyr - OMe	Tyr - Leu - NH_2
Phe - Tyr	Tyr - Leu
Tyr - Tyr - NH_2	Ac - Tyr - Phe - NH_2
Ac - Tyr - OEt	Tyr - Phe - NH_2
Tyr - NH_2	diZ - Tyr - Thr - OMe
	Tyr - Thr - NH_2

Several of the peptides in Table II are sparingly soluble. This
may have obscured digestion that might have occurred had the concentrations been higher. Series of peptides containing Leu or Phe
residues have also been examined. Seventy peptides have been shown
to resist cleavage by cathepsin D.

    It may be concluded that cathepsin D is quite similar to pep-
sin in its substrate specificity, two adjacent hydrophobic groups
providing the most susceptible linkage, and rates of hydrolysis increasing markedly as peptide chain length grows up to about six
residues. Cathepsin D, however, has a much more restricted series
of di- and tri-peptides that it can hydrolyze, and must have obvious
differences in its binding site.

    An interesting series of small peptides is represented by mem-
bers such as Ala-Phe-NH_2, His-Phe-NH_2, Phe-Phe-NH_2, and Phe-Phe.
These peptides were all cleaved by cathepsin D preparations at
purities of 50-100 units/mg protein, but when the purity rose to
150-200 units/mg these activities were lost. This digestion was
obviously due to an impurity, possibly an aminopeptidase contaminant.
The odd feature was that cleavage of these peptides was completely
blocked by the addition of pepstatin or by the enzyme preparation's
reaction with diazoacetylnorleucine methyl ester in the presence of
cupric ions. These treatments also abolished all activity against
the peptides listed in Table I. It must be concluded that the
peptidase impurity had all the characteristics of a carboxyl type
of enzyme. Possibly, some of the multiple forms of cathepsin D had
a broader specificity than others, and these were lost as purification progressed.

BIOLOGICAL ROLE OF CATHEPSIN D

Tritosome study

    To explore the role of cathepsin D, it is necessary to know
something of its ability to digest proteins within the digestive
vacuolar system of the cell. This is because cathepsin D has little
action at pH's greater than 5.5-6.0, so most theories of its action
postulate that activity is confined to the lysosomes where the pH
may fall below 5.0 (13), or to regions immediately at the cell surface as in the osteoclast (14). Studies of the digestive capacity
of cathepsin D and other lysosomal cathepsins have been conducted
by several workers including Coffey and de Duve (15) and Huisman
et al. (16). In most such studies, however, the pH has been arbi-
trarily fixed at 5.0. Together with Dr. John R. Lewis, I have
undertaken a study of the digestion of two proteins over an extended
range of pH. The proteins are human serum albumin and bovine hemoglobin, poor and good substrates, respectively, for cathepsin D at
pH 5.

We prepared tritosomes by the usual methods (17), injecting Triton WR-1339 in rats, removing the livers four days later, and preparing tritosomes by density gradient centrifugation. These preparations were then ruptured to release their cathepsins and added to substrate preparations at various pH values (Fig.2). At each pH, a series of incubations was performed with various additions: (1) dithiothreitol and EDTA, (2) iodoacetate, (3) pepstatin, DTT, and EDTA, (4) pepstatin and iodoacetate. The first tube would be expected to reflect the full proteolytic activity of all proteinases except for metalloproteases; the second would be expected to permit only the expression of non-thiol proteases; the third would be expected to show no cathepsin D activity, and the fourth would be expected to block all thiol and carboxyl enzymes. Calculations based on these four values were used to estimate the results in Figure 2.

It can be seen from Figure 2 that albumin digestion is due almost entirely to thiol proteases over the pH range 3.5-6.5. Pepstatin-inhibitable activity (cathepsin D) makes a modest contribution of about 20% of the total digestion in the region around pH 4.5. With hemoglobin, cathepsin D accounts for almost all of the digestion at pH 3 and for about 40% of the activity at pH 4.5-5.5. The residual activity after thiol proteinases and cathepsin D have been inhibited is neglible in albumin digestion, but there is some non-inhibitable activity accounting for 5-10% of the digestion of hemoglobin. These findings essentially agree with those of Dean (18), who concludes that cathepsin D and thiol proteinase together account for almost all of the general proteolytic activity of the tritosome.

In the next sections I shall consider some examples of biological processes in which cathepsin D seems to play a role, as judged by elevations in its activity and by other evidence pointing to the involvement of the lysosomal digestion mechanism within the cell.

Embryonic development

Some details have been presented on the changes of cathepsin D in the development of the embryonic chick femur (9). The peak level of enzymatic activity occurs at about the 15th day of development, at the same time the bone is undergoing extensive change or remodeling. The appearance of the collagen breakdown product hydroxyproline is maximal at this time, as is the rate of calcium deposition. Two breakdown processes are presumably in full swing at this time, proteoglycan and matrix degradation are occurring in advance of the moving calcification zone and the central marrow space is being enlarged at the expense of the surrounding bone. These events are all more intense in the shaft of the femur and become attenuated toward the ends of the bond.

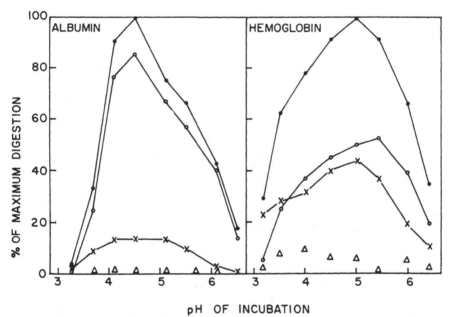

Figure 2. The digestion of human serum albumin and bovine hemoglobin by disrupted rat liver tritosomes. ● - total digestion in the presence of dithiothreitol and EDTA; o- iodoacetate-inhibitable protease thiol proteinase; x - pepstatin-inhibitable protease (cathepsin D); Δ - protease resistant to pepstatin and iodoacetate.

Another embryonic system of interest is the developing skin. In this tissue there is a striking breakdown of the structural components of the skin that appear to be related to a rapid enlargement of the bases of the feather shafts (19). This is most pronounced on the 13th day of incubation. Figure 3 illustrates the change in the level of free proline at this time and shows the parallel change in cathepsin D. The most obvious explanation for these findings is that a rapid degradation of protein is taking place in the remodeling skin, and the lysosomal system, including cathepsin D, is actively involved in the process.

Uterine involution

The rat uterus presents one of the most extreme cases of tissue resorption. In a period of 96 hr following the birth of the litter, the mother's uterus decreased by 85% in weight, muscle mass, and collagen content (2). A study of cathepsin D behavior during this period shows a continuing, almost linear increase, in enzymic activity during the entire 96 hr of involution (Fig.4). When involution is essentially complete, there is a slow return of cathepsin D activity to a base-line level. This return is about half-complete by ten days post partum (2). Clearly much of the cathepsin D activity of the uterus is lysosomal, but quantitative measurements are difficult because the organ is tough and fibrous. The data are interpreted as pointing to an important role of lysosomal proteases in the breakdown of the uterine tissue. Electron microscopic studies indicate that these lysosomes may be in both muscle and macrophage-like cells (20). Both cell types are conserved in involution, leading to increasing concentrations of lysosomal enzymes as the surrounding tissues diminish. The total amount of cathepsin D in the uterus as a whole is maximal at parturition, so that conservation alone could account for the increasing concentration of enzyme (2). Continued synthesis of the enzyme is most likely occurring during this period.

Cartilage breakdown

Early searches for the cause of proteoglycans loss from the matrix of cartilage revealed prominent proteolytic activity in the cartilage at pH 3.5-5 (21). This activity was eventually shown to be that of cathepsin D, the single-most prominent proteolytic enzyme in the cartilage. It has been demonstrated in several laboratories that cathepsin D degrades the proteoglycan structure quite rapidly with a pH optimum between 4 and 5 (22,23). The enzyme has been purified extensively from various types of cartilage, including rabbit ear, chick embryo limb, and human articular cartilage (23,24). These preparations display the typical characteristics of cathepsin D with regard to molecular weight, substrate specificity, and

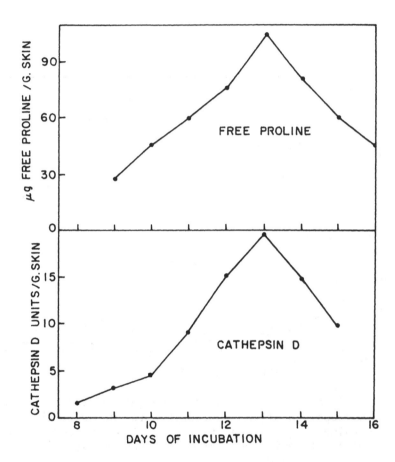

Figure 3. Changes in free proline and cathepsin D in the skin removed from the backs of chick embryos after various periods of incubation.

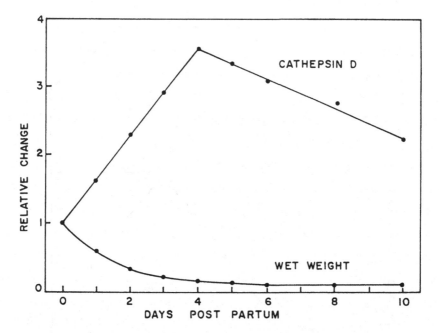

Figure 4. Changes in total uterine weight and the concentration of cathepsin D of the post-partum rat uterus. The
relative value of 1 is equal to 2.7 g wet weight and 45 units
of cathepsin D (μequivalent Tyr released/min/g wet tissue).

existence of multiple forms. Upon extensive purification, however, the enzyme lost all ability to degrade proteoglycans at pH values greater than 5.5-6.0 (23). Therefore, the role of this protease in the cartilage remains problematic.

There is a clear rise of cathepsin D activity in degenerative diseases such as osteoarthritis. The enzyme activity in discolored regions of human articular cartilage on the margins of osteoarthritic lesions is 3.4 units/g tissue, whereas the activity in normal white regions of the cartilage is 1.5 units/g (25). Similar results have been reported by Shoji and Granda (26). The pH limitations, however, seem to require that the enzyme function inside chondrocyte lysosomes or immediately at the cell surface. Since cells are widely spaced in the cartilage matrix, it does not seem that this enzyme could be of major importance in the widespread destruction of the matrix.

Prompted by these considerations, I have been conducting a survey, with Dr. Asher Sapolsky, of possible neutral proteases in the cartilage. As part of this study human articular cartilage was homogenized, extracted and fractionated on DEAE-Sephadex columns. Several peaks of neutral proteolytic activity against proteoglycan were recovered (27). There was a generous admixture of cathepsin D in these fractions. Therefore, to obtain a pH profile of the neutral proteases over the range 4-9, it was necessary to add pepstatin to prevent any interference by cathepsin D. We observed, in addition to a neutral protease peak at pH 7.3, two additional peaks at pH 4.5 and 5.5 that failed to be inhibited by pepstatin or by thiol reagents. They were, however, blocked by chelating agents. Cartilage apparently contains two acid metalloproteases. Perhaps they are related to the neutral metalloproteases that are the current center of our attention (27). They promise to be of considerable interest as members of a class of acid proteases that is virtually unknown.

ACKNOWLEDGMENT

The author is indebted to several colleagues mentioned in the text who assisted in certain phases of this work. Appreciation is due to Mrs. Carolyn Taplin and Dr. Rodolfo Denis for their competent technical assistance. This research was supported in part by NIH grants AM-16540 and HD-06773.

SUMMARY

The purity of cathepsin D has been increased from 150 units/mg to over 200 units/mg. Peptides such as Ala-Phe-NH_2, His-Phe-NH_2 and Phe-Phe were split by impure enzyme and activity was blocked by pepstatin and diazoacetylnorleucine methyl ester. Pure preparations no

longer digested these peptides. This points to the presence of a
second peptidase activity similar to cathepsin D in specificity and
inhibition properties, but distinct from it. Cathepsin D splits the
peptides Leu-Phe-NH$_2$, Leu-Tyr-NH$_2$, Ac-Phe-TyrI$_2$, and Ala-Leu-Tyr-Leu
upon overnight incubation. More rapid splitting is found with phenyl
sulfite, Glu-Ala-Leu-Tyr-Leu-Val, and Bz-Arg-Gly-Phe-Phe-Leu-4-
methoxy-β-naphthylamide.

Digestion of bovine hemoglobin and human serum albumin by
ruptured rat liver tritosomes was studied over the pH range 2.5-6.5.
The combined action of cathepsin D and thiol proteinases accounted
for most of the digestion. Cathepsin D accounted for 75% of the
hemoglobin digestion at pH 3 and 45% at pH 5. Thiol proteinase
accounted for 85% of the albumin digestion at pH 5. The role of
cathepsin D in the development of embryonic limbs and skin, in
uterine involution, and in cartilage degradation was reviewed. The
activity of cathepsin D on cartilage matrix proteoglycans is limited
to acid pH values. Human articular cartilage also contains metallo-
proteases active at pH 4.5 and 5.7.

REFERENCES

1. Press, E. M., Porter, R. R., and Cebra, J. (1960) Biochem. J.
    74, 501-514
2. Woessner, J. F., Jr. (1965) Biochem. J. 97, 855-867
3. Shamberger, R. J., Jr. (1963) Studies on Uterine Cathepsins,
    Doctoral Dissertation, University of Miami, pp.77
4. Woessner, J. F., Jr., and Shamberger, R. J., Jr. (1971) J. Biol.
    Chem. 246, 1951-1960
5. Sapolsky, A. I., and Woessner, J. F., Jr. (1972) J. Biol. Chem.
    247, 2069-2076
6. Barrett, A. J. (1970) Biochem. J. 117, 601-607
7. Ferguson, J. B., Andrews, J. R., Voynick, I. M., and Fruton,
    J. S. (1973) J. Biol. Chem. 248, 6701-6708
8. Keilova, H., and Keil, B. (1968) Collect. Czech. Chem. Commun.
    33, 131-140
9. Woessner, J. F., Jr. (1971) in Tissue Proteinases (Barrett, A.J.
    and Dingle, J. T., eds) pp. 291-308, North Holland Publ. Co.,
    Amsterdam
10. Smith, R. E. (1975) in Lysosomes in Biology and Pathology
    (Dingle, J. T., and Dean, R. T., eds) Vol. IV, pp. 193-249,
    North Holland Publ. Co., Amsterdam
11. Matyash, L. P., Belyaev, S. V., Oksenoit, E. S., and Stepanov,
    V. M. (1974) Biokhimiya 39, 197-200
12. Sampath-Kumar, P. S., and Fruton, J. S. (1974) Proc. Natl. Acad.
    Sci. U.S.A. 71, 1070-1072
13. Jensen, M. S., and Bainton, D. F. (1973) J. Cell. Biol. 56,
    379-388

14. Vaes, G. (1968) J. Cell Biol. 39, 676-697
15. Coffey, J. W., and de Duve, C. (1968) J. Biol. Chem. 243, 3255-3263
16. Huisman, W., Lanting, L., Doddema, H. J., Bouma, J. W. M., and Gruber, M. (1974) Biochim. Biophys. Acta 370, 297-307
17. Leighton, F., Poole, B., Beaufay, H., Baudhuin, P., Coffey, J. W., Fowler, S., and de Duve, C. (1968) J. Cell Biol. 37, 482-513
18. Dean, R. T. (1976) Biochem. Biophys. Res. Commun. 68, 518-523
19. Woessner, J. F., Jr. (1970) in Chemistry and Biology of the Intercellular Matrix (Balasz, E. A., ed) Vol. III, pp. 1663-1669 Academic Press, Inc., New York
20. Parakkal, P. F. (1969) J. Cell Biol. 41, 345-354
21. Lucy, J. A., Dingle, J. T., and Fell, H. B. (1961) Biochem. J. 79, 500-508
22. Weston, P. D., Barrett, A. J., and Dingle, J. T. (1969) Nature 222, 285-286
23. Woessner, J. F., Jr. (1973) J. Biol. Chem. 248, 1634-1642
24. Sapolsky, A. I., Howell, D. S., and Woessner, J. F., Jr. (1974) J. Clin. Invest. 53, 1044-1053
25. Sapolsky, A. E., Altman, R. D., Woessner, J. F., Jr., and Howell, D. S. (1973) J. Clin. Invest. 52, 624-633
26. Shoji, H., and Granda, J. L. (1974) Clin. Orthop. Relat. Res. 99, 293-297
27. Sapolsky, A. I., Howell, D. S., and Woessner, J. F., Jr. (1976) J. Clin. Invest. 58, 1030-1042

ACID PROTEASE AND ITS PROENZYME
FROM HUMAN SEMINAL PLASMA

Pintip Ruenwongsa and Montri Chulavatnatol

Department of Biochemistry, Faculty of Science
Mahidol University, Bangkok, Thailand

An acid protease with an optimum pH of 2.5 was first described in human seminal plasma as pepsin and pepsinogen (1), but had not been purified or characterized. Recently, we have purified the acid protease and its proenzyme from human seminal plasma (2,3). In many respects, the properties of seminal plasma acid protease are similar to those of gastric pepsin. Since the proenzyme is more stable than the active enzyme in alkaline solution and can be converted into its active form in acidic solution, the acid protease is likely to exist in seminal plasma, at the physiological pH around 7.5 (4), in proenzyme form.

It seems improbable for the proenzyme to be activated in the slightly alkaline pH of seminal plasma. Under physiological conditions, however, when the semen is freshly deposited in the vagina and mixed with the vaginal discharge, a slightly acidic environment is created (5). The pH of vaginal fluid is changed from 4.3 to 7.2, and from 3.5 to 5.5, in normal and low fertility respectively, after mixing with semen. Therefore, it is of particular interest to investigate the acid dependency of activation of the proenzyme.

It has been suggested that the proteolytic activity of seminal plasma is an essential prerequisite to sperm penetration through the cervical mucus (6); thus, the possible physiological function of the acid protease is also discussed in this report.

EXPERIMENTAL PROCEDURE

Purification of Acid Protease

The proenzyme and active form of acid protease were purified as described previously (3).

Assays for Acid Protease Activity

The activity of both proenzyme and active acid protease were assayed with acid-denatured hemoglobin substrate by the method of Kassell and Meitner (7); the proenzyme is converted to the active form in the acid conditions of the assay. The 2 ml assay mixture contained 0.1 M citrate-phosphate buffer, pH 2.5, 25 mg hemoglobin and appropriately diluted enzyme concentration. The assays were performed for 30 min at 37° and terminated by addition of 2 ml of 5% trichloroacetic acid; the absorbance at 280 nm of soluble fraction was then measured. To assay proenzyme, the pH of the incubation was raised to 8 with 1 M Tris to destroy the active form. This solution was then acidified to activate and assay for the proenzyme.

Activation of Proenzyme of Acid Protease

The proenzyme solution was incubated at 37° in 0.1 M citrate-phosphate buffer at the desired acid pH. After an appropriate time interval, aliquots were transferred into 1 M Tris, pH 8.0 and further incubated for 30 min, stopping activation as well as destroying the active enzyme. The remaining proenzyme was then assayed at pH 2.5 as described.

Hydrolysis of Cervical Mucus Protein by Acid Protease

The cervical mucus protein, containing 1-2 mg protein, was incubated with 5 mg of acid protease in 0.1 M citrate-phosphate buffer, pH 2.5 at 37° for 1 hr. After stopping the reaction by raising the pH of the incubating mixture to 7 with NaOH, it was then lyophilized. Digestion of the protein was detected by using SDS-gel electrophoresis (8).

Protein Determination

Protein concentration was determined by the method of Lowry et al. (9) using bovine serum albumin as standard.

RESULTS

Acid protease activity in seminal plasma at pH 3.0 and 7.5. Previous results (3) have shown that the acid protease, pH optimum of

2.5, is unstable at pH 7.5, the physiological pH of human seminal
plasma. Thus, an experiment on pH stability of an acid protease in
human seminal plasma was performed. As shown in Table I, the proteolytic activity found before and after adjusting seminal plasma to
pH 3 appeared to be the same; but when the latter was readjusted to
its original pH of 7.5, its activity was lost. Since the proenzyme
of the acid protease was stable at pH 7.5, this finding clearly indicated that the acid protease occurred in seminal plasma as a proenzyme which could be converted into its active form when the pH of
seminal plasma was lowered to 3. In addition, the active enzyme was
destroyed when pH of seminal plasma was raised to 7.5. This was
supported by result shown in Figure 1; the protease activity was
completely destroyed within 90 sec after exposure to pH 8.

Conversion of the proenzyme into its active form. We concluded
that only the proenzyme form of acid protease can exist in human
seminal plasma at the physiological pH of around 7.5; therefore, we
investigated activation of proenzyme in acid medium. The purified
proenzyme was incubated in 1 mM HCl, pH 3, for 1 hr, and then chromatographed on Sephadex G-50 column. The proenzyme was converted into
an active form and some peptide of small molecular weight was released (Fig.2). As shown in Table II, when the amino acid analyses of
the proenzyme, the active form, and activation peptide were carried
out, the number of each amino acid residue of the proenzyme agreed
well with the additive value between the number of that amino acid
in the active form and activation peptide. This supported the conversion of the proenzyme to an active form. The amino acid composition of the active protease and of the proenzyme were comparable to
those of bovine pepsin (10) and pepsinogen (11). However, definite
differences are present. About forty residues which carried most of
the basic amino acids, were released from pepsin, while sixty-nine
residues, which carried about 30% of basic amino acid of the precursor
were liberated from the proenzyme of seminal plasma acid protease.

pH-dependency of the proenzyme activation. As shown in Figure 3,
both the rate and extent of the proenzyme activation were dependent
on pH. The activation was essentially completed after 10 min at pH 2
and 3; but more of the proenzyme remained at 30 min when the activation was performed at higher pH value, from 4 to 5. At these high
pH values, the percentage of proenzyme remaining did not change when
the activation continued for more than 30 min. At the end of 30 min,
if the pH of the activation was brought down to 2, the activation
would immediately accelerate to a completion.

Between pH 5 and 2, as the pH was lowered, the rate of activation increased. For activation at pH 2, 3 and 4, if the remaining
proenzyme were plotted semilogarithmically against incubation time,
a linear line would be observed, suggesting a first-order reaction.

Table I

Acid protease activity in human seminal plasma

The activity of acid protease in seminal plasma was determined at 37° for 30 min using hemoglobin as substrate. 2 ml of reaction mixture contains 15 mg Hb, 20 λ seminal plasma (corresponding to 0.6 - 0.7 enzyme unit) and 0.1 M citric acid-phosphate buffer, pH 2.5.

Treatment of Seminal Plasma Prior to Assay	Acid Protease Activity (ΔA_{280}/30 min)
None	0.65
None, 37°/2 hr	0.66
Adjusted to pH 3, 37°/1 hr*	0.66
Readjusted to pH 7.5, 37°/1 hr**	0.05

*Seminal plasma, pH 7.5, was adjusted to pH 3 with 4 N HCl and incubated at 37° for 1 hr before assaying.

**The pH 3 treated-seminal plasma was readjusted to pH 7.5 with 4 N NaOH and incubated at 37° for 1 hr before assaying.

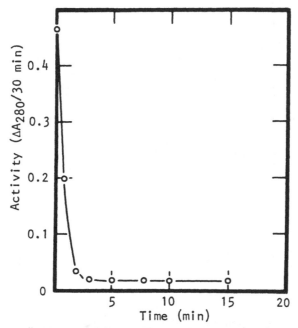

Figure 1. Stability of the active acid protease at pH 8.
The enzyme was incubated in 0.1 M sodium phosphate buffer,
pH 8 at 37°. Aliquots were removed at different time
intervals and assayed for proteolytic activity at pH 2.5,
using hemoglobin as substrate.

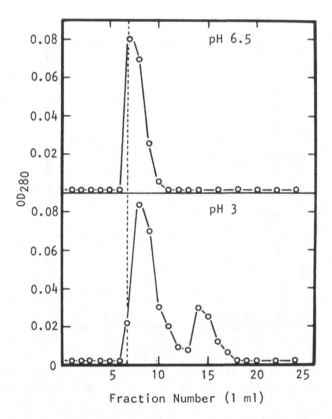

Figure 2. Gel filtrations on Sephadex G-50 (0.9 x 50 cm):
(Top) proenzyme in 0.02 M sodium phosphate buffer, pH 6.5.
(Bottom) proenzyme was incubated in 1 mM HCl at 37° for
1 hr and then chromatographed in 1 mM HCl, pH 3.

Table II

Amino acid composition of the proenzyme, the acid protease, and the activated peptide of human seminal plasma.

Each sample containing 5 nmoles of protein was hydrolyzed in a sealed and evacuated tube at 110° for 22 hr with 6 N HCl. The analysis was performed with a Beckman automatic amino acid analyzer.

Amino acid	Proenzyme[a]	Activation peptide	Acid protease	Human pepsin (15)
		residues/mole		
Aspartic acid	38 (36)	6	30	40
Glutamic acid	55 (53)	10	43	31
Threonine	29 (30)	4	26	27
Serine	35 (37)	6	31	43
Proline	23 (23)	4	19	19
Glycine	41 (43)	6	37	35
Alanine	26 (26)	5	21	18
Valine	27 (25)	3	24	27
Half cystine	8 (6)	1	5	6
Methionine	4 (4)	0	4	5
Isoleucine	15 (15)	3	12	25
Leucine	34 (35)	6	29	22
Tyrosine	20 (20)	3	17	15
Phenylalanine	19 (20)	3	17	15
Tryptophan	N.D.[b]	N.D.[b]	N.D.[b]	5
Lysine	15 (14)	4	10	0
Histidine	4 (4)	1	3	1
Arginine	11 (10)	4	6	3
Total	406 (404)	69	335	337

[a] Number in parenthesis is the sum of the values found in the acid protease and in the activation peptide.

[b] N.D. = not determined

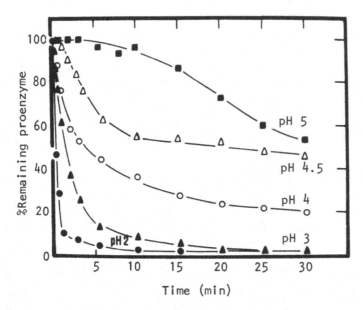

Figure 3. The time course of activation of the proenzyme at various pH values. The proenzyme (8µg/ml) was incubated in 0.1 M citrate-phosphate buffer of the desired pH. The proenzyme remaining at various time intervals was assayed as described in "Experimental Procedure".

At pH 4.5 and 5, there was an initial lag period followed by a more
rapid decline in enzyme concentration, suggesting a deviation from a
first order to a more complex reaction. A detailed discussion of
this activation mechanism appears in a separate paper (12) and in
chapter 6 of this book.

Hydrolysis of protein substrate by acid protease at various pH.
Protein substrates other than hemoglobin were digested at pH values
higher than its optimum pH. The results obtained in Table III were
in agreement with those observed in pH profile (2); i.e. that none
of the proteins could be hydrolyzed at pH above 4.5.

Hydrolysis of cervical mucus protein by the acid protease. The
pattern of cervical mucus protein before and after hydrolyzing by
acid protease were shown in Figure 4. The major protein band was
lost in cervical mucus treated with acid protease. The appearance
of new bands of smaller molecular weight indicated that hydrolysis
had occurred. However, this protein band was also lost in cervical
mucus hydrolyzed with chymotrypsin at pH 7.8, suggesting a rather
nonspecific hydrolysis. The cervical mucus protein seemed to be
hydrolyzed to a greater extent by the seminal plasma acid protease
than by chymotrypsin.

DISCUSSION

Clearly, the proenzyme is the only form of acid protease that
exists in human seminal plasma at its physiological pH of 7.2-7.8.
If, on the other hand, the proenzyme form could be converted into
its active form in seminal plasma, the latter would have then been
destroyed at the slightly alkaline pH of seminal plasma. Thus, the
protease activity would not have remained constant, but would have
decreased as the time of incubation increased. Since this was not
observed, the activation must not be taking place in the seminal
plasma.

Upon activation in acid, the proenzyme converts into the active
form probably by a limited proteolysis, which results in the release
of a peptide, containing sixty-nine amino acid residues (Table II).
Our other results (12) suggested that it seems to involve a single
cleavage of a specific peptide bond in the proenzyme molecule,
releasing only one activation peptide. In contrast, activation of
bovine gastric zymogen, pepsinogen (13) and prochymosin (14) produce
a multiplicity of peptides by the cleavage of 7-8 bonds. However,
cleavage of only one of these bonds seems to be required to release
active enzyme, and the other 6-7 bonds may be susceptible linkage
sensitive to proteolysis in general. It should be noted that the
amino acid composition of acid protease from human seminal plasma
(335 residues) compares favorably with that of human pepsin (337
residues) (15).

Table III
Digestion of bovine serum albumin, ovalbumin and dimethylcasein
at various pH by acid protease from seminal plasma

The reaction mixtures contained 2µg enzyme, 15 mg protein
substrate and 0.1 M citric acid-sodium phosphate buffer in
a total volume of 2 ml. The assays were performed at 37°
for 30 min.

pH	Protein Substrates		
	Bovine Serum Albumin	Ovalbumin	Dimethylcasein
	$\Delta A280/30$ min		
2.5	0.06	0.04	0.35
3.5	0.01	0.01	0.10
4.5	0	0	0.05
5.5	0	0	0
6.5	0	0	0

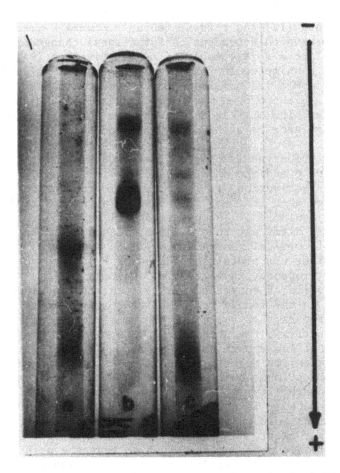

Figure 4. Polyacrylamide gel electrophoresis in 0.1% SDS
of; a) acid protease-hydrolyzed cervical mucus in citric-
phosphate buffer, pH 2.5; b) non-hydrolyzed cervical mucus;
and c) chymotrypsin-hydrolyzed cervical mucus in phosphate
buffer, pH 7.8. The electrophoresis was performed, using
7% gel, at the current of 8 ma per tube. Proteins were
stained with coomassie brilliant blue and destained in
acetic:methanol:H$_2$0 (75:50:875).

The activation process of the proenzyme is a complicated reaction which is influenced by the pH of the medium. The kinetics
of activation shift from the first order between pH 2 and 4 to more
complexity with a lag period between pH 4.5 and 5 (Fig.3). Nevertheless, the mechanism of activation involving at least two steps
has been proposed (12) and represented as P $\rightleftharpoons$ P' $\rightleftharpoons$ A + t.
The first step is a pH-dependent conformational change of the native
proenzyme (P) into a form capable of self-activation (P'). The
second step is a unimolecular hydrolytic release of an activation
peptide (t) from P'; the remainder of P' becomes the active acid
protease (A). This proposed mechanism can explain the incompleteness
of activation at high pH (Fig.3) in that the equilibrium of the
second step may favor P' at high pH value and shift to A at low pH.

Since the proenzyme is progressively converted into its active
form at pH below 6 and the pH of vaginal fluid is fairly acidic to
bring about the activation, it is conceivable that the proenzyme
can be activated when it comes into contact with the acidic vaginal
fluid during ejaculation. The physiological role of seminal plasma
acid protease is difficult to elucidate, although it can hydrolyze
the proteins of cervical mucus (Fig.4). The hydrolysis seems to be
nonspecific, since the proteins can also be hydrolyzed by seminal
plasma neutral protease, chymotrypsin, and trypsin (6). Moreover,
when the pH profile of enzyme activity, stability, activation as
well as the pH in the female reproductive tract are considered
together, it appears that the pH in the female reproductive tract is
low enough to activate the proenzyme and to maintain the enzyme
activity, but not low enough to allow proteolytic activity. On the
other hand, the enzyme might function at physiological pH higher than
its in vitro pH optimum, but the result in Table III seems not to
support this possibility. Extensive evidence is available, however,
that the acid protease is not the only enzyme involved in reproduc-
tion that requires acid pH for its action (16,17). Thus, the
possibility might exist that the acid environment might somehow
reside in the female reproductive tract so that the acid protease
could exert its activity towards certain protein substrates.

ACKNOWLEDGMENT

This study was supported by the Rockefeller Foundation and Funds
for Overseas Research Grants and Education, Inc.

SUMMARY

An acid protease, with an optimum pH of 2.5, exists in seminal
plasma in a proenzyme form. In acidic pH, the proenzyme is converted
into the active form, resulting in the release of small molecular

weight peptide. The extent and rate of proenzyme-active enzyme conversion is absolutely dependent on pH. Between pH 5 and 2, as the pH is lowered, the extent of conversion increases and reaches a maximum between pH 3 and 2. The kinetics of activation shift from first-order between pH 2 and 4 to more complexity with a lag period between pH 4.5 and 5. Under physiological conditions, the proenzyme might be activated by coming into contact with acidic vaginal fluid during ejaculation. The acid protease can hydrolyze the cervical mucus protein.

REFERENCES

1. Lundquist, F., and Seedorff, H. H. (1952) Nature, 170, 1115-1116
2. Ruenwongsa, P., and Chulavatnatol, M. (1974) Biochem. Biophys. Res. Commun. 59, 44-50
3. Ruenwongsa, P., and Chulavatnatol, M. (1975) J. Biol. Chem. 250, 7574-7578
4. Mann, T. (1964) The Biochemistry of Semen and of the Male Reproductive Tract, pp. 88, Methuen, London
5. Fox, C. A., Meldrum, S. J., and Watson, B. W. (1973) J. Reprod. Fertil. 33, 69-75
6. Moghissi, K. S., and Syner, F. N. (1970) Int. J. Fertil. 15, 43-49
7. Kassell, B., and Meitner, P. A. (1970) Methods Enzymol. 19, 337-347
8. Weber, K., and Osborn, M. (1969) J. Biol. Chem. 244, 4406-4412
9. Lowry, O. H., Rosebrough, N. J., Farr, A. L., and Randall, R. J. (1951) J. Biol. Chem. 193, 265-275
10. Helga, M., Lang, S. S. J., and Kassell, B. (1971) Biochem. 10, 2296-2301
11. Chow, R. B., and Kassell, B. (1968) J. Biol. Chem. 243, 1718-1724
12. Chulavatnatol, M., and Ruenwongsa, P. (1976) Biochim. Biophys. Acta 452, 525-532
13. Harboe, M., Anderson, P. M., Foltmann, B., Kay, J., and Kassell, B. (1974) J. Biol. Chem. 249, 4487-4494
14. Foltmann, B., and Andersen, B. (1971) FEBS Lett. 17, 87-89
15. Mills, J. N., and Tang, J. (1967) J. Biol. Chem. 242, 3093-3097
16. Zaneveld, L. J. D., Polakoski, K. L., and Schumacher, G. E. B. (1973) J. Biol. Chem. 248, 564-570
17. Zaneveld, L. J. D., Polakoski, K. L., and Williams, W. L. (1973) Biol. Reprod. 9, 219-225

LIST OF COMMUNICATIONS AND POSTERS

Communications

B. M. Dunn, W. G. Moesching, C. Deyrup, M. L. Trach, R. J. Nolan, and W. A. Gilbert, University of Florida, Gainesville, "Synthesis of Pepsin Inhibitor".

M. S. Silver, Amherst College, Amherst, Massachusetts, "Do Pepsin-Catalyzed Hydrolyses Ever Give Rise to Amino-Enzyme Intermediates?".

H. Nakatani, Kyoto University, Kyoto, Japan, "Zinc (II)-PAD Complex as a Catalytic Site Probe for Acid Proteases".

E. Hackenthal and R. Hackenthal, University of Heidelberg, Heidelberg, FRG, "Rat Brain Isorenin: Comparison with Renal Renin".

J. Burton, K. Poulsen, and E. Haber, Harvard Medical School, Boston, "The Design of Peptides Which are Competitive Inhibitors of Renin".

V. Turk, I. Kregar, F. Gubensek, T. Turneek, I. Urh and M. Kovacic, University of Ljubljana, Ljubljana, Yugoslavia, "Some Biochemical Studies on Native and Immobilized Cathepsin D and Cathepsin S".

S. Roffman, S. Lerner and L. M. Greenbaum, Columbia University, New York, "pH-Dependent Activity of Cathepsin D from Neoplastic Cell Using a Sensitive (H^3)-Hemoglobin Substrate".

M. R. Levy, S. C. Chou and W. A. Siddiqui, Southern Illinois University and University Hawaii, "Protease Inhibitors and Growth of the Malarial Parasite".

K. Morihara, T. Oka, K. Oda and S. Murao, Shionogi Col, and University of Osaka Prefecture, "Specificity of Microbial Acid Proteinases on Synthetic Substrates, Relationship to Their Behavior Against Pepstatin".

Posters

I. Rubin, E. Lauritzen and M. Lauritzen, University of Copenhagen, Denmark, "Purification of Renin from Rat Kidney by Affinity Chromatography Using the Substrate Analogy NH_2-Leu-Leu-Val-Tyr-Ser-COOH".

R. T. Dworschack and M. P. Printz, University of California at San Diego, "Covalently-Immobilized Ligands as Probes of the Pepstatin Binding Site on Renin".

H. C. Li and N. Back, State University of New York, Buffalo, New York, "An Acid Protease Kinin-Forming System from Fibroblast Cell Line L-929".

I. M. Samloff and V. Dadufalza, UCLA School of Medicine, Los Angeles, "The Effect of pH and Ionic Strength on the Proteolytic Activities of Human Group I and Group II Pepsins".

A. M. Faerch and B. Foltmann, University of Copenhagen, Denmark, "Mucor miehei Protease, Selected Sequences".

V. B. Pedersen and B. Foltmann, University of Copenhagen, Denmark, "Comparison of the Activation of Pepsinogen and Prochymosin".

R. Delaney, University of Oklahoma Health Sciences Center, Oklahoma City, "Three Molecular Forms of Rhizopus chinensis Acid Protease".

P. Sepulveda and J. Tang, Oklahoma Medical Research Foundation, Oklahoma City, "Partial Sequences of Human Pepsin and Gastricsin".

J. S. Huang and J. Tang, Oklahoma Medical Research Foundation, Oklahoma City, "Large-Scale Purification of Cathepsin D from Porcine Spleen".

J. S. Huang and J. Tang, Oklahoma Medical Research Foundation, Oklahoma City, "Purification of Microbial Acid Proteases by Pepstatin-Sepharose Affinity Chromatography".

M. K. Harboe, Institute of Biochemical Genetics, University of Copenhagen, Denmark, "Bovine Pepsinogen B and Pepsin B".

D. H. Rich, E. Sun and J. Singh, University of Wisconsin, Madison, "Synthesis of Dideoxy-Pepstatin, Mechanism of Inhibition of Porcine Pepsin".

CONFERENCE PARTICIPANTS

Andreeva, Natalia
Institute of Molecular Biology, Academy of Sciences of U.S.S.R.
Moscow, U.S.S.R.   117312

Antonov, Vladimir K.
Shemyakin Institute of Bioorganic Chemistry, U.S.S.R. Academy of
Sciences, 117312 Moscow B-312 ul. Vavilova 32, Moscow, U.S.S.R.

Back, Nathan
State University of New York at Buffalo, Bell Facility, Room B-107
180 Race Street, Buffalo, New York 14207

Barrett, Alan J.
Strangeways Research Laboratory, Worts' Causeway
Cambridge, England CB1 4RN

Basha, S. M. Mahaboob
Department of Biochemistry, JHM Health Center, University of Florida
Box J245, Gainesville, Florida 32610

Basu, S. P.
Oklahoma Medical Research Foundation, 825 N.E. 13th
Oklahoma City, Oklahoma 73104

Blundell, Tom
Birbeck College, London University, Malet Street, London, UK WC1E7HX

Bowers, William E.
The Rockefeller University, 1230 York Avenue, New York, New York 10021

Bradford, Reagan H.
Oklahoma Medical Research Foundation, 825 N.E. 13th
Oklahoma City, Oklahoma 73104

Burton, James
Cardiac Unit, Massachusetts General Hospital, Boston, MA 02114

Capréol, Suzanne
University of Waterloo, Waterloo, Ontario, Canada

Davies, David
National Institutes of Health, Building 2, Room 316
Bethesda, Maryland 20014

Delaney, Robert
Department of Biochemistry, Oklahoma University Health Sciences Center
Oklahoma City, Oklahoma 73104

Dunn, Ben M.
Department of Biochemistry and Molecular Biology, JHM Health Center
University of Florida, Box J245, Gainesville, Florida 32610

Dworschack, Robert
Department of Medicine M-013, University of Calif., San Diego
LaJolla, California 92093

Esmon, Naomi
Oklahoma Medical Research Foundation, 825 N.E. 13th
Oklahoma City, Oklahoma 73104

Faerch, Anne-Marie
Institute of Biochemical Genetics, University of Copenhagen
ØFarimagsgade 2A, Copenhagen K DK-1353, Denmark

Foltmann, Bent
Institute of Biochemical Genetics, University of Copenhagen
ØFarimagsgade 2A, Copenhagen K DK 1353, Denmark

Hackenthal, Eberhard
Department of Pharmacology, University of Heidelberg
Im Nevenheimer Feld 366, 69 Heidelberg, Germany

Harboe, Marianne
Hansen's Laboratory, Sct. Anne Plads, Copenhagen 1250, Denmark

Hartsuck, Jean
Oklahoma Medical Research Foundation, 825 N.E. 13th
Oklahoma City, Oklahoma 73104

Hofmann, Theo
Department of Biochemistry, Medical Science Building
University of Toronto, Toronto, Ontario M5S 1A8, Canada

Holzer, Helmut
Biochemistry Biochemisches Institut der Universitat
Hermann-Herder-Strasse 7, D-7800 Freiburg im Breisgau, Germany

Hsu, I-Nan
Department of Biochemistry, University of Toronto, Toronto M5S 1A8
Canada

Huang, Jung San
Oklahoma Medical Research Foundation, 825 N.E. 13th
Oklahoma City, Oklahoma 73104

Huang, Shuan
Oklahoma Medical Research Foundation, 825 N.E. 13th
Oklahoma City, Oklahoma 73104

Inagami, Tadashi
Department of Biochemistry, School of Medicine, Vanderbilt University
Nashville, Tennessee 37232

Jackson, Kenneth W.
Oklahoma Medical Research Foundation, 825 N.E. 13th
Oklahoma City, Oklahoma 73104

James, Michael N. G.
Department of Biochemistry, University of Alberta,
Edmonton, Alberta T6G 2H7, Canada

Jih, Mike
Oklahoma Medical Research Foundation, 825 N.E. 13th
Oklahoma City, Oklahoma 73104

Johnson, B. Connor
Oklahoma Medical Research Foundation, 825 N.E. 13th
Oklahoma City, Oklahoma 73104

Kaizer, Emil Thomas
Department of Chemistry, University of Chicago, 5735 S. Ellis Avenue
Chicago, Illinois 60637

Kassell, Beatrice
Department of Biochemistry, Medical College of Wisconsin
Milwaukee, Wisconsin 53233

Kay, John
Department of Biochemistry, University College, P. O. Box 78
Cardiff CF1 1X2, Wales, UK

Kollmorgen, G. Mark
Oklahoma Medical Research Foundation, 825 N.E. 13th
Oklahoma City, Oklahoma 73104

Lanier, Paul
Oklahoma Medical Research Foundation, 825 N.E. 13th
Oklahoma City, Oklahoma 73104

Leckie, Brenda J.
Blood Pressure Unit, Western Infirmary, Glasgow G12 6NT, UK

Lee, Diana M.
Oklahoma Medical Research Foundation, 825 N.E. 13th
Oklahoma City, Oklahoma 73104

Levy, Michael R.
Southern Illinois University, Edwardsville, Illinois 62026

Lin, Tsau-Yen
Merck Institute for Therapeutic Research, Rahway, New Jersey 07065

Liu, Diane
Oklahoma Medical Research Foundation, 825 N.E. 13th
Oklahoma City, Oklahoma 73104

Liu, Mamie
National Institutes of Health, Building 2, Room 314
Bethesda, Maryland 20014

Marchand, Alan
Department of Chemistry, University of Oklahoma, 620 Parrington Oval
Norman, Oklahoma 73069

Marciniszyn, Joseph, Jr.
Oklahoma Medical Research Foundation, 825 N.E. 13th
Oklahoma City, Oklahoma 73104

McPhie, Peter
N.I.A.M.D.D., Building 10, Room 9N 119, Bethesda, Maryland 20014

Morihara, Kazuyuki
Shionogi Research Laboratory, Shionogi and Co., Ltd.
5-12-4, Sagisu, Fukushima-ku, Osaka, Japan 553

Nakatani, Hiroshi
Faculty of Agriculture, Kyoto University, Kyoto 606, Japan

Oda, Kohei
Roswell Park Memorial Institute, 666 Elm Street
Buffalo, New York 14263

Peanasky, Robert J.
Department of Biochemistry, School of Medicine
The University of South Dakota, Vermillion, South Dakota 57069

Pedersen, Vibeke Barkholt
Institute of Biochemical Genetics, University of Copenhagen
Øfarimagsgade 2A, Copenhagen K DK 1353, Denmark

Pratt, John
Oklahoma Medical Research Foundation, 825 N.E. 13th
Oklahoma City, Oklahoma 73104

Rao, S. Narasinga
Oklahoma Medical Research Foundation, 825 N.E. 13th
Oklahoma City, Oklahoma 73104

Rich, Daniel H.
University of Wisconsin, 425 N. Charter Street
Madison, Wisconsin 53706

Rickert, William S.
Department of Statistics, University of Waterloo
Waterloo, Ontario N2L 3G1, Canada

Roffman, Steven
Department of Pharmacology, College of Physicians and Surgeons
Columbia University, 630 West 168, New York, New York 10032

Rubin, Inger
Department of Biochemistry A, University of Copenhagen
Blegdamsvej 3, DK-2200 Copenhagen N., Denmark

Ruenwongsa, Pintip
Department of Pharmacology, School of Medicine, Yale University
333 Cedar Street, New Haven, Connecticut 06510

Samloff, I. Michael
Chief, Division of Gastroenterology, Harbor General Hospital
1000 West Carson Street, Torrance, California 90509

Schneider, Larry W.
Oklahoma Medical Research Foundation, 825 N.E. 13th
Oklahoma City, Oklahoma 73104

Sepulveda, Patricia
Oklahoma Medical Research Foundation, 825 N.E. 13th
Oklahoma City, Oklahoma 73104

Silver, Marc S.
Amherst College, Amherst, Massachusetts 01002

Smith, Robert E.
Lawrence Livermore Laboratory, University of California
Bio-Med Division L-523, Livermore, California 94550

Subramanian, E.
National Institutes of Health, N.I.A.M.D.D., Building 2, Room 312
Bethesda, Maryland 20014

Sugai, Setuko
Faculty of Agriculture, Kyoto University, Kyoto, Japan 606

Tang, Jordan
Protein Studies Laboratory, Oklahoma Medical Research Foundation
825 N.E. 13th, Oklahoma City, Oklahoma 73104

Turk, Vito
Department of Biochemistry, J. Stefan Institute
University of Ljubljana, 61001 Ljubljana, Jamova 39
P. O. Box 199, Ljubljana, Yugoslavia

White, Clayton S.
Oklahoma Medical Research Foundation, 825 N.E. 13th
Oklahoma City, Oklahoma 73104

Woessner, J. Frederick, Jr.
Professor of Biochemistry, School of Medicine, University of Miami
P. O. Box 520875, Miami, Florida 33152

Alaupovic, Petar
Lipoprotein Laboratory, Oklahoma Medical Research Foundation
825 N.E. 13th, Oklahoma City, Oklahoma 73104

McConathy, Walter J.
Lipoprotein Laboratory, Oklahoma Medical Research Foundation
825 N.E. 13th, Oklahoma City, Oklahoma 73104

Olofsson, Sven
Lipoprotein Laboratory, Oklahoma Medical Research Foundation
825 N.E. 13th, Oklahoma City, Oklahoma 73104

Vitto, Anthony
Biology Division, Oak Ridge National Laboratory
Oak Ridge, Tennessee 37830

Hanson, Livy
Veterans Administration Hospital, 921 N. E. 13th
Oklahoma City, Oklahoma 73104

Guia, Marcos Mares
Department of Biochemistry, Federal University
Delo Horizonte, Minas Gerais, Brazil

Printed in the United States
by Baker & Taylor Publisher Services